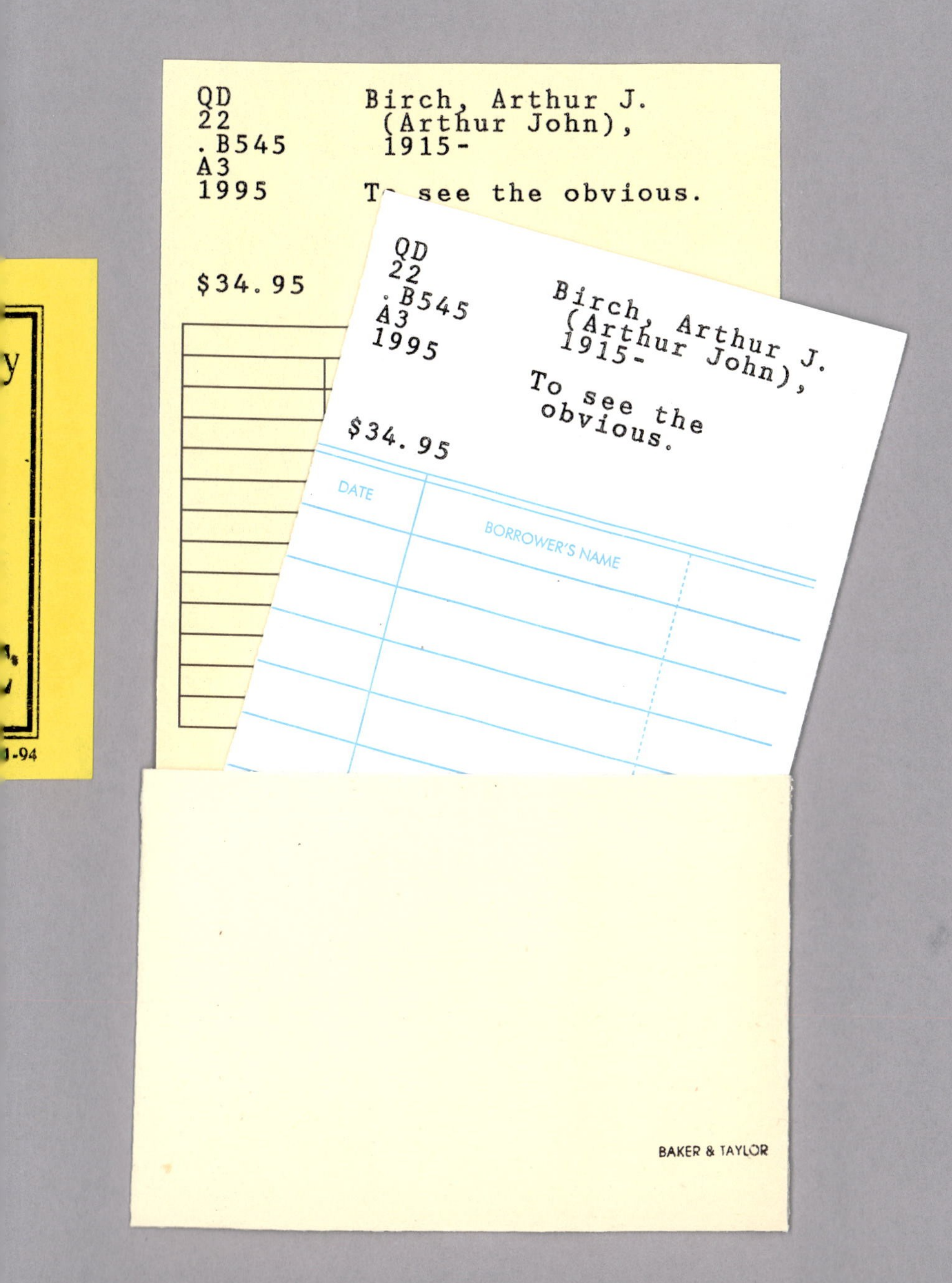
QD
22
.B545
A3
1995
Birch, Arthur J.
(Arthur John),
1915-
To see the obvious.
$34.95
DATE
BORROWER'S NAME
BAKER & TAYLOR

To See the Obvious

To See the Obvious

Arthur J. Birch

PROFILES, PATHWAYS, AND DREAMS
Autobiographies of Eminent Chemists

Jeffrey I. Seeman, Series Editor

American Chemical Society, Washington, DC 1995

Library of Congress Cataloging-in-Publication Data

Birch, Arthur J. (Arthur John), 1915–
To see the obvious / Arthur J. Birch.

p. cm. — (Profiles, pathways, and dreams, ISSN 1047–8329)

Includes bibliographical references and index.

ISBN 0–8412–1840–4 (case)

1. Birch, Arthur J. (Arthur John), 1915–
2. Chemists—Australia—Biography. 3. Chemistry, Organic—Australia—History—20th century.

I. Title. II. Series.

QD22.B545A3 1995
540′.92—dc20
[B]

95–18601
CIP

Jeffrey I. Seeman, Series Editor

The paper used in this publication meets the minimum requirements of American National Standard for Information Sciences—Permanence of Paper for Printed Library Materials, ANSI Z39.48–1984.

(∞)

PRINTED IN THE UNITED STATES OF AMERICA

Profiles, Pathways, and Dreams

Foreword

In 1986, the ACS Books Department accepted for publication a collection of autobiographies of organic chemists, to be published in a single volume. However, the authors were much more prolific than the project's editor, Jeffrey I. Seeman, had anticipated, and under his guidance and encouragement, the project took on a life of its own. The original volume evolved into 22 volumes, and the first volume of Profiles, Pathways, and Dreams: Autobiographies of Eminent Chemists was published in 1990. Unlike the original volume, the series was structured to include chemical scientists in all specialties, not just organic chemistry. Our hope is that those who know the authors will be confirmed in their admiration for them, and that those who do not know them will find these eminent scientists a source of inspiration and encouragement, not only in any scientific endeavors, but also in life.

Contributors

We thank the following corporations and Herchel Smith for their generous financial support of the series Profiles, Pathways, and Dreams.

Akzo nv

Bachem Inc.

DuPont

Duphar B.V.

Eisai Co., Ltd.

Fujisawa Pharmaceutical Co., Ltd.

Hoechst Celanese Corporation

Imperial Chemical Industries PLC

Kao Corporation

Mitsui Petrochemical Industries, Ltd.

The NutraSweet Company

Organon International B.V.

Pergamon Press PLC

Pfizer Inc.

Philip Morris

Quest International

Sandoz Pharmaceuticals Corporation

Sankyo Company, Ltd.

Schering–Plough Corporation

Shionogi Research Laboratories, Shionogi & Co., Ltd.

Herchel Smith

Suntory Institute for Bioorganic Research

Takasago International Corporation

Takeda Chemical Industries, Ltd.

Unilever Research U.S., Inc.

Profiles, Pathways, and Dreams

Titles in This Series

Sir Derek H. R. Barton *Some Recollections of Gap Jumping*

Arthur J. Birch *To See the Obvious*

Melvin Calvin *Following the Trail of Light: A Scientific Odyssey*

Donald J. Cram *From Design to Discovery*

Michael J. S. Dewar *A Semiempirical Life*

Carl Djerassi *Steroids Made It Possible*

Ernest L. Eliel *From Cologne to Chapel Hill*

Egbert Havinga *Enjoying Organic Chemistry, 1927–1987*

Rolf Huisgen *The Adventure Playground of Mechanisms and Novel Reactions*

William S. Johnson *A Fifty-Year Love Affair with Organic Chemistry*

Raymond U. Lemieux *Explorations with Sugars: How Sweet It Was*

Herman Mark *From Small Organic Molecules to Large: A Century of Progress*

Bruce Merrifield *Life During a Golden Age of Peptide Chemistry: The Concept and Development of Solid-Phase Peptide Synthesis*

Teruaki Mukaiyama *To Catch the Interesting While Running*

Koji Nakanishi *A Wandering Natural Products Chemist*

Tetsuo Nozoe *Seventy Years in Organic Chemistry*

Vladimir Prelog *My 132 Semesters of Chemistry Studies*

John D. Roberts *The Right Place at the Right Time*

Paul von Rague Schleyer *From the Ivy League into the Honey Pot*

F. Gordon A. Stone *Leaving No Stone Unturned: Pathways in Organometallic Chemistry*

Andrew Streitwieser, Jr. *A Lifetime of Synergy with Theory and Experiment*

Cheves Walling *Fifty Years of Free Radicals*

About the Editor

JEFFREY I. SEEMAN received his B.S. with high honors in 1967 from the Stevens Institute of Technology in Hoboken, New Jersey, and his Ph.D. in organic chemistry in 1971 from the University of California, Berkeley. Following a two-year staff fellowship at the Laboratory of Chemical Physics of the National Institutes of Health in Bethesda, Maryland, he joined the Philip Morris Research Center in Richmond, Virginia. In 1983–1984, he enjoyed a sabbatical year at the Dyson Perrins Laboratory in Oxford, England, and claims to have visited more than 90% of the castles in England, Wales, and Scotland.

Seeman's 90 published papers include research and patents in the areas of photochemistry, nicotine and tobacco alkaloid chemistry and synthesis, conformational analysis, pyrolysis chemistry, organotransition metal chemistry, the use of cyclodextrins for chiral recognition, and structure–activity relationships in olfaction. He was a plenary lecturer at the Eighth IUPAC Conference on Physical Organic Chemistry held in Tokyo in 1986 and has been an invited lecturer at numerous scientific meetings and universities. Currently, Seeman serves on the Petroleum Research Fund Advisory Board. He continues to count Nero Wolfe and Archie Goodwin among his best friends.

Contents

Photographs

Preface

"HOW DID YOU GET THE IDEA—and the good fortune—to convince 22 world-famous chemists to write their autobiographies?" This question has been asked of me, in these or similar words, frequently over the past several years. I hope to explain in this preface how the project came about, how the contributors were chosen, what the editorial ground rules were, what was the editorial context in which these scientists wrote their stories, and the answers to related issues. Furthermore, several authors specifically requested that the project's boundary conditions be known.

As I was preparing an article[1] for *Chemical Reviews* on the Curtin–Hammett principle, I became interested in the people who did the work and the human side of the scientific developments. I am a chemist, and I also have a deep appreciation of history, especially in the sense of individual accomplishments. Readers' responses to the historical section of that review encouraged me to take an active interest in the history of chemistry. The concept for Profiles, Pathways, and Dreams resulted from that interest.

My goal for Profiles was to document the development of modern organic chemistry by having individual chemists discuss their roles in this development. Authors were not chosen to represent my choice of the world's "best" organic chemists, as one might choose the "baseball all-star team of the century". Such an attempt would be foolish: Even the selection committees for the Nobel prizes do not make their decisions on such a premise.

The selection criteria were numerous. Each individual had to have made seminal contributions to organic chemistry over a multidecade career. (The average age of the authors is over 70!) Profiles would represent scientists born and professionally productive in different countries. (Chemistry in 13 countries is detailed.) Taken together, these individuals were to have conducted research in nearly all subspecialties of organic chemistry. Invitations to contribute were based on solicited advice and on recommendations of chemists from five continents, including nearly all of the contributors. The final assemblage was selected entirely and exclusively by me. Not all who were invited chose to participate, and not all who should have been invited could be asked.

A very detailed four-page document was sent to the contributors, in which they were informed that the objectives of the series were

1. to delineate the overall scientific development of organic chemistry during the past 30–40 years, a period during which this field has dramatically changed and matured;

2. to describe the development of specific areas of organic chemistry; to highlight the crucial discoveries and to examine the impact they have had on the continuing development in the field;

3. to focus attention on the research of some of the seminal contributors to organic chemistry; to indicate how their research programs progressed over a 20–40-year period; and

4. to provide a documented source for individuals interested in the hows and whys of the development of modern organic chemistry.

One noted scientist explained his refusal to contribute a volume by saying, in part, that "it is extraordinarily difficult to write in good taste about oneself. Only if one can manage a humorous and light touch does it come off well. Naturally, I would like to place my work in what I consider its true scientific perspective, but . . ."

Each autobiography reflects the author's science, his lifestyle, and the style of his research. Naturally, the volumes are not uniform, although each author attempted to follow the guidelines. "To write in good taste" was not an objective of the series. On the contrary, the authors were specifically requested not to write a review article of their field, but to detail their own research accomplishments. To the extent that this instruction was followed and the result is not "in good taste", then these are criticisms that I, as editor, must bear, not the writer.

As in any project, I have a few regrets. It is truly sad that Egbert Havinga and Herman Mark, who each wrote a volume, and David Ginsburg, who translated another, died during the course of this project. There have been many rewards, some of which are documented in my personal account of this project, entitled "Extracting the Essence: Adventures of an Editor" published in *CHEMTECH.*[2]

Acknowledgments

I join the entire scientific community in offering each author unbounded thanks. I thank their families and their secretaries for their contributions. Furthermore, I thank numerous chemists for reading and reviewing the autobiographies, for lending photographs, for sharing information, and for providing each of the authors and me the encouragement to proceed in a project that was far more costly in time and energy than any of us had anticipated.

I thank my employer, Philip Morris USA, and J. Charles, R. N. Ferguson, K. Houghton, H. Grubbs, and W. F. Kuhn, for without their support Profiles, Pathways, and Dreams could not have been. I thank ACS Books, and in particular, Cheryl Wurzbacher (production manager), Janet Dodd (senior editor), Joan Comstock (department head), and their staff for their hard work, dedication, and support. Each reader no doubt joins me in thanking 24 corporations and Herchel Smith for financial support for the project.

I thank my children, Jonathan and Brooke, for their patience and understanding; remarkably, I have been working on Profiles for more than half of their lives—probably the only half that they can remember! Finally, I again thank all those mentioned and especially my family, friends, colleagues, and the 22 authors for allowing me to share this experience with them.

JEFFREY I. SEEMAN
Philip Morris Research Center
Richmond, VA 23234

April 19, 1993

[1] Seeman, J. I. *Chem. Rev.* **1983,** *83,* 83–134.

[2] Seeman, J. I. *CHEMTECH* **1990,** *20*(2), 86–90.

Editor's Note

ARTHUR BIRCH has been a scientist at heart from his earliest years, years that now span nearly 8 decades. Who could have forecast that a young, independent whippersnapper of a child in mid-1930s Australia, called "Professor"by his grandmother, would provide a heritage of such magnitude in science and education? He had to leave Australia to obtain his Ph.D.; none was granted in his home country at that time! He has played a major role in Australian chemistry, first in Sydney and later in building the Research School of Chemistry at the Australian National University in Canberra (ANU). Birch's education connects Oxford, with Sir Robert Robinson, and Cambridge, with Lord Todd, with his own professorial positions at Manchester, Sydney, and ANU. Robinson himself had been at Manchester (twice!), as well as Sydney.

Studying for his doctorate in philosophy at Oxford, working with Robinson and others at the Dyson Perrins Laboratory, discovering the Birch Reduction, and living within the Oxford spires and magic must have been a heady mix for a young scientist. The lifelong influence of Cambridge, but even more especially of Oxford, on Birch are reflected in this book and in some of his historical and philosophical publications.

According to Sir Derek Barton, whose career overlapped that of his friend, it was Birch's creativeness and inventiveness combined with his strength of character and ability to understand what is important that drove Birch to study dissolving metal reductions. "I found the key paper on dihydrobenzene in 1942," recounts Birch. "I told Robinson about it personally when I presented my first paper for him to add his name (which he did not!). He knew [of the work]; he did not approve. I simply told him that I interpreted his expressed lack of interest as permission to carry on this work independently! I was still working on his projects; I just worked twice as hard. [Years later, in a radio] broadcast, Robinson mentioned the reduction work as a highlight of his time in Oxford."

Barton explained, "Robinson was away from Oxford during much of the war, and when he found out about Birch's work, he was furious." Robinson was a consultant for ICI and Birch was an ICI employee on a government grant at Oxford. When Birch showed his first reduction manuscript to ICI, as was required by terms of the fellowship, ICI initially refused publication as they were independently working on similar chemistry. "I went to see them and they gave way," Birch recalls. "ICI was angry with me and probably with Robinson as well, for he was supposed to be supervising my work. I continued over Robinson's objections because I knew it was the most important thing to do." Barton concluded, "Birch was certainly very seriously considered for a Nobel Prize, but I think that Sir Robert torpedoed it! He shot [Sir Christopher] Ingold's chances for a Nobel Prize, as well. Robert Robinson acted on his animosities."

Robinson described Arthur Birch as the student most like himself. In turn, Birch recently wrote of Robinson, "He often dreamed and did so in a highly effective subconscious way; he could see the 'obvious'." *To See the Obvious* . . . a telling title for Birch's autobiography, reinforcing the ties between these two giants. The intensity of his years with Robinson is such that Robinson seems never far from Birch. I recall a number of times with Birch; it was as if Robinson were with us as well. As Birch recalled, "Robinson and I had a 'friendly enemy' relationship. In a strange way, I was almost his father. He was a childish fellow; I tried to keep him in good humor." It has been his fate that their lives are indelibly intertwined.

Chemists everywhere have come to know the "Birch Reduction" for its versatility and reliability. Twenty five years ago, the Birch Reduction became one of my sacred cows, a dependable friend that would amazingly convert readily obtainable aromatic compounds to key intermediates for my graduate research project. Only later did I learn that the Birch Reduction was also used commercially by Syntex and other pharmaceutical companies on a unfathomable scale—"kilos and kilos of metal and liquid ammonia," recalled Carl Djerassi recently. "Birch was 10 years ahead of his time in three areas: reduction chemistry, biosynthesis, and organometallics," adds Barton. For many, including myself, the Birch Reduction is in the same category as the Robinson Annelation and the Diels–Alder Reaction.

Carl Djerassi, a renowned chemist and author in his own right, characterizes Birch as "a maverick, a lone wolf." "He always has been very imaginative, highly creative. We picked him as a Syntex consultant. Some of my non-Syntex colleagues asked, 'Why bring a fellow all the way from Australia to Mexico City.' 'It's worth the few more dollars if you have hired a real brain,' I answered."

"I'm a private person; that is, I am not an extrovert," Birch explains. "I do what my conscience tells me I must, whatever the personal sacrifice. I find that I fulfill the definition of an eccentric—with the exception that I can spell. I decided in my middle age that I am a rather strange person. I don't behave conventionally. I'm somewhat shy. I'm reluctant to put forward my personal views unless I am compelled to do so. I am also a perfectionist. It is a bit of a problem to want everything as perfect as possible. I have the ability; I feel that I also have the responsibility. I have never believed in virtue generated solely by self-sacrifice.

Outside the Dyson Perrins Laboratory, Oxford, 1946. Sir Robert Robinson and Lady Robinson are seated prominently in the center of the first row. "I am in the second row, first on the left. There are many well-known faces, a high proportion alas now dead," according to Birch. From left to right: first row, (second) J. A. Barltrop, (fourth) J. C. Smith, (fifth) S. G. P. Plant, (sixth) E. Hope, (ninth) D. L. Hammick; second row, (fifth) Renee Jaeger, (eighth) M. T. Rogers (origin of the title of this book), and my Australian colleagues, (ninth) John W. Cornforth, and (twelfth) Rita H. Cornforth; third row (last on right) R. L. Huang, VC of Hong Kong and member of government.

"A few months ago," Birch recalled, "I was interviewed for an article in Chemistry in Australia.* I really don't like the article. It is entitled 'A Man on His Own' because I never was dependent on collaboration for strategic ideas during my scientific career. I have had several extremely capable colleagues who helped develop my ideas. Herchel Smith was one of these. I do not have broad, indiscriminate friendships. Most of my close friendships are notably with my family. I received two beautiful cards from two of my grandchildren when I was in hospital last week."

Birch certainly is proud of his Australian heritage and his contributions to Australian education. "I remember being with Birch in Tasmania," recalled his long-time friend Koji Nakanishi recently. "We walked around and Birch's reminiscences were quite memorable."

"Birch likes to pretend that he is a tough Australian," summarizes Barton. "He deals with people quite frankly, often using colorful language. He came from a poor family, pulled himself up by his remarkable ability and hard work, always had a scholarship. Birch learned how to work hard and continues to do so."

Arthur Birch has spent nearly 10 years working on this book. During the early drafts in the late 1980s, I offered numerous suggestions to improve the readability and organization of the material. Birch's response: "I am sending you my final personal version (which I am still trying to reorder) with numbered paragraphs. What could help, at cost of your labour, is this. A list of paragraph numbers in the order you would like them. I guarantee to accept this." As I already had an ASCII version of the manuscript, I carefully studied the text and sent Birch not a listing of numbers but a revised manuscript, keeping all of the material but reordering it. Shortly thereafter, Birch reported that he had begun to rewrite the entire book. What you have in your hands now is the culmination of many versions and many years, the result of enormous energy on Birch's part—energy he maintains to this day!

We have been aiming to have this volume appear by Arthur's 80th birthday, August 3, 1995. On May 25, 1995, I received a fax from Australia that provided additional impetus to meet that goal. Jessie Birch, Arthur's wife, wrote, "Arthur is in hospital—renal failure that may prove to be untreatable. He is having ultrasound this morning, but the picture is not good. In any event, the time span remains short. Under the circumstances, please do what you can to finish the book as quickly as possible. [You can] make final decisions." I immediately called Margaret Brown, ACS Books Senior Production Specialist, and we agreed to proceed as rapidly as possible, without making the final additions and modifications. I reassured Jessie Birch that we would honor her request.

One week later, June 2nd, I called Canberra. Birch himself answered the phone! His voice was very weak, but distinctly Arthur. As he tells it, "I just

*Wright, B. A. *A Man on His Own, Chemistry in Australia.* March 1995, 34–38. In the article, Birch was quoted as saying "Nobody in 10 years has asked me for an opinion on anything." Readers of this volume may chuckle as did I when reading this quote! Upon reading the final draft of this essay, "Jessie chuckled over your summary, so I take that as approval," reported Birch by fax. "I am very flattered by your sketch, although I seem rather a 'wierdo.' Perhaps I am!"

refused to die. I'm very strong minded. Probably too strong minded." Birch then reported that he had been working on the last set of page proof corrections I had requested 10 days earlier and that he would soon fax four pages of information. Birch, just dismissed from the hospital, clearly was going to finish his autobiography, and do it right! He most certainly was not going to sit on the sidelines . . . that's Arthur Birch.

Tradition, love of knowledge, continuity, and pride are some of the traits of Oxford and Cambridge. I cherish these feelings from my own sabbatical year at Oxford in 1983–1984. Perhaps nothing illustrates this better than the annual laboratory photographs, illustrated by the 1946 Oxford photograph on page xxv and the 1950 Cambridge photograph on page xxviii. I can barely walk past the decades of photographs on the third floor of the Dyson Perrins Laboratory without stopping to gaze and reflect. The two reproduced here are lovely because they pull together so many people who were central to Arthur Birch's life: Robinson and Todd, the Cornforths, and Herchel Smith, among others.

JEFFREY I. SEEMAN
Philip Morris Research Center
Richmond, VA 23234

June 15, 1995

Silhouette by S. John Ross

Research workers and the academic staff in organic, inorganic, and theoretical chemistry at Cambridge University in 1950. "I am in the second row, third from the left. Herchel Smith is also in the second row, second from the right. Many English professors of the 1970s and 1980s came from this group. From left to right: first row, (ninth) Professor H. J. Emeleus, (tenth) Professor A. R. Todd (later Lord Todd), (eleventh) Professor Sir J. E. Lennard-Jones. Gordon Stone, a contributor to Profiles, Pathways, and Dreams is standing fifth from the right in the fourth row; on his left is H. G. Khorana. Next to me is Barbara Thornber, then Todd's efficient secretary and later Mrs. F. G. Mann; Dr. F. G. Mann is to Professor Lennard-Jones's right in the front row," according to Birch.

To See the Obvious

Arthur J Birch

Prolog

THIS IS THE PERSONAL STORY, over 50 years, of the making of one chemist, embracing general social, political, ethical, and historical aspects of scientific research of technical and philosophic significance, applied in organization and teaching. My "hero" is organic chemistry in its adolescence up to its present adult status.

I look, from my experience, at the sociology of creative science with the motivations, rewards, and characteristics of creative scientists. I recount the origins and courses of some intellectual advances and social and technical applications of science due to chance and to design. I illustrate my theme by means of specific cases that directly involved me over 50 years, so that I know exactly what happened and why. At the age of 79—all passion spent—I assert that I have no personal reservations or interests except to record for the future.

You are the best person I know for seeing the obvious.

M. A. T. Rogers (of ICI) to me in 1946

Rogers's unintended high compliment marks my theme. What, during the past 50 years of organic chemistry research, was "obvious" and why? Why does one person perceive it and others not? What are the appropriate ways to develop perceptions? What were the obvious opportunities and problems to be seen against the various technical backgrounds and the needs of changing times? How is the obvious to be related to chemistry as a whole, with ideas from one area beneficially influencing those of other apparently unrelated areas? What attitudes, training, and emotions make a successful scientist? What are the rewards and problems? My personal experiences can provide an illustration.

Why Me?

Why should a 10-year-old boy, in the mid-1920s in isolated Sydney, Australia, decide to become a scientist? I had no models in my local family (mostly Australian farmers) and never met a scientist until I went to the university. My decision was not a conscious choice of a career, about which I knew nothing. I had no idea that my life was already affected in many practical ways by the results of science. Rather, I had a burning desire to know all about everything and was prepared to work to acquire this knowledge. All children are born with a faculty of wonder, and even at 79 my "cloud of glory" has not quite faded into the light of common day.

How Science Works

Science seems to have an almost organic life of its own, with the growing points in its continuously accumulating complex of ideas and methods attested by parallel and simultaneous individual approaches. At certain periods specific developments become obvious or "ripe". Individuals do affect rates of progress on a micro scale, but I agree with Michel de Montaigne, who said in the late 16th Century, "He who is mounted highest has often more honor than merit, for he is got up but an inch upon the shoulders of the last but one." A better known aphorism by Newton, but going back to Bernard of Chartres in the 12th Century, involves "standing on the shoulders of giants" (but not *only* giants!).

The human tendency is to individualize various achievements, but a mass of rather anonymously striving individuals forms the necessary base for advances within the scientific Zeitgeist. Some people stand out in popular esteem for various reasons that are not always scientifically creditable; they may be famous for being famous. All of the workers should accrue their own personal esteem as necessary contributors to a great human endeavor. The difficulty in awarding individual Nobel Prizes (and frequently the injustice of it) is an index of the problem. Nobel Prizes are supposed to be awarded for contributions to human welfare, not to science. Montaigne's dictum stands; the next person in line is often awarded that prize.

Gaps in the Record

If I do not adequately acknowledge my scientific predecessors here, it is because this is only one story of personal insights from one participant among the many necessary to build an authentic historical picture of any period within its boundaries of knowledge and understanding. The exact precursors of ideas are difficult to define in any case. Chemistry includes very mixed-up situations with parallel developments by different individuals, each seeing the obvious from a different position in the complex of discovery, invention, social capabilities, available resources, and definable needs of the time. The few genuine breakthrough ideas are often difficult to discern when they appear.

Difficulties abound in dealing retrospectively with the record. Precursor ideas may have been expressed too soon; they may have been forgotten, at least at the conscious level, or be inaccessible to later workers. Abstractors and indexers are fallible; reviewers usually only update. Even a survey of the theoretically complete record may be misleading as to what was genuinely influential in the progress of ideas. Those who participated at the time, who have good memories and who noticed, can alone give one authentic story of a development.

I use several criteria for evaluating the influence of an idea. One sign is that its implications must have been recognized by its author; without that a discovery or invention becomes merely an observation. I recall congratulating Robinson on his 1930s suggestion of the origin of cholesterol from squalene. He laughed, "I have had six other ideas since then; they can't all be right, can they?" The cholesterol theory had in fact no influence and was resurrected as a curiosity. To exercise due influence, publication of an idea must be timely, which is a problem. It is appropriate to be a year or two ahead of the broadly obvious, but ideas that are 10 or more years ahead of their time usually lapse and have to be regenerated independently.

Personalities

Why discuss my personal history when my subject is organic chemistry? Although science is the investigation of what is there, the obvious becomes revealed through sets of personal interactions with Nature. To see this process as a human activity requires an understanding of the makeup of researchers, their psychologies, their emotional and other rewards, their socialization, their differing viewpoints, and their dedication. Explicit personal stories may also be of use to stimulate embryonic scientists to look at themselves and their scientific styles, to show them how science works as a personal endeavor rather than a set of mechanical advances.

Friedrich Auguste Kekulé (1829–1896) summed up the pursuit of creative science thus: "Let us learn to dream, there do we find the Truth, but let us be silent until our vision has been put to the Truth by our awakened understanding." He could have included more specific reference to the hard work required in putting comprehensions to the test. Here I recount some of my dreams and the results of putting them to the test.

Prelude, and Evolution of a Chemist

The details of a research career are dictated partly by accident, but largely by inclinations and temperament.

Family Background: My Father

I discuss my mother's critical influence on my career later. I was an only child. My father, Arthur Spencer Birch, was a sceptical Englishman born in Daventry, Northamptonshire, to a father who was master of the workhouse, District Registrar in Northampton, and organist and choirmaster in the beautiful "Perpendicular" Church in Kettering. The first half of his 10 children had a good education, but my father was in the second half. After quarrels with his increasingly drunken father he left home at 14 and began supporting himself as a "snotty-bit picker" in a Northampton boot factory. (His job was to sort the leather scraps on which the workers spat, hence the name.) He was very intelligent, although not formally educated. He left school at 12 and could never spell properly. His eldest sister, Margaret, was one of the early suffragettes. She married a Canadian (Lewis) and became inspector of factories in Alberta. (Her birth centenary was recently celebrated.) My father joined her but could not stand the Canadian winters. He was on the Yukon–Alaskan Boundary Survey (1908–1910), worked in logging camps, and canoed down the Athabasca River. When he wanted to go places, he worked on ships. He went to Fiji for several years to warm up. Then he took part in a gold rush in New Zealand, where he met my mother, Lily Bailey (and his fate, married at 37).

Arthur Spencer Birch, my father, a pastry chef who refused to eat his own beautiful confections.

Somehow my father had learned to cook. He was pastry chef at the major hotel in Sydney (Wentworth), making fancy wedding cakes, carving swans in ice, and so forth. He was artistic; I still have some of his drawings. He cooked for Nellie Melba, whom he heartily disliked, and later became manager of Woolworth's Cafeterias.

His entire family was interesting. One brother, Donald, a clergyman who lost his faith, became mayor and schoolmaster of a town of Scandinavian immigrants in Canada, which he ruled with a rod of iron for 40 years. A sister, Jessie, was dietician to the London County Council; another brother, John, became secretary of the Baptist Union of South Africa. Others were last heard of in the United States. My father became terminally ill when I was 14 and died of heart failure when I was 21. He was an atheist to the end, despite experiments with unorthodox religions like theosophy.

I acquired several major attitudes from my father. One was a curiosity about the whole world and an inclination not to take anything for granted. The other attitude was related to the lost independence of his youth. He worked very hard to support his family, but clearly what mattered to him was his lost autonomy, his inability as an employee to make his own decisions. I grew up feeling that autonomy was the most important thing to seek, and that wealth and position were only contributors to that end. Alas, in my selfish teenage years I understood him little, a matter for great regret: "The moving finger writes and having writ moves on...." (E. Fitzgerald's translation of Omar Khayyam).

Primary School

I did not attend primary school until I was 7 (the compulsory age). This relatively late start may explain to some extent my early lack of socialization. I had already learned to read widely, I cannot recall how, and I regarded my first school reading lessons, based on simple phonetics, as ridiculous. This attitude immediately led to a habit of being skeptical about what I was told at school. In my first class I astonished the teacher by spelling "beautiful", but then I read "echo" phonetically and protested to her about the idiocy of English pronunciation. She was not pleased to hear it from a 7-year-old.

I sauntered carelessly through primary school. The teaching must have been good, if conventional. My only effort was to keep about level with a special friend, at about 3rd to 6th from the top in the class. I was eventually promoted a year in one step, which made me eligible at 10 to take the permit to enroll (entrance exam) for high school. My parents considered me too young, so I endured a year of boring business principles, shorthand, and bookkeeping, all of which I have forgotten (consciously, anyhow).

My classroom then was next to the science area, which was set off by glazed partitions. I used to watch what was going on there instead of paying attention to what I was supposed to be doing. The science teacher (who had a slight stammer) eventually asked me to give some lectures. He said, "One day you will be a professor." Thus I acquired a taste for science at about 10. I recall being very intrigued about why a stick standing in water appeared bent, and being fascinated when he explained the reason.

My last primary teacher, Margaret Mundy, caught me reading under my desk. My rude reply was, "I find the lesson so boring, I am reading about bridge-building instead." I still have a letter of hers, from a school in Goulburn in 1933: "... I was very pleased to see your name in the paper, I hope you enjoy the work in High School and will be as successful as you were in the 7th grade ... no one but you obtained the High School pass..." She went on to say, "I am finding girls vastly different to boys. Instead of having them impatient to give an answer as many of you were, one has to beg them even to think..."

My beloved Aunt Maude, who died in 1924 with a letter from me under her pillow, left me a legacy of 100 pounds (quite a lot then). The inheritance enabled my father to buy me some basic chemical equipment, including a Liebig condenser (after trouble with Customs), a Kipps generator, and some books. One of my first experiments was making alcohol from figs growing in our garden and converting it into ether and ethyl bromide. I taught myself organic chemistry from Cohen's textbook (from about the age of 12). I still have it, although it is slightly battered.

My hidden depths appeared through a curious incident. The headmaster of the Woollahra Superior Public School, who disliked rowdy boys, forbade us to enter the school grounds before assembly. This meek, disciplined boy (me) joined with other ringleaders to organize what would now be called a sit-in to protest the order. I was hauled up before the threatening and blustering headmaster (even then it seemed to me he was scared), caned, and sent back to my class. The master, named "Beery" for good reasons, met me with a sweet smile. "I

About age 16 in front of the garden fig tree. The alcohol for some of my first syntheses (ethyl bromide and diethyl ether) came from fermenting the abundant figs.

am surprised at you, Birch," he said, but I gathered that his surprise was not derogatory. He just wondered why a well-behaved boy like me had become involved. I do not like being pushed around by idiots, and I learned then that positive action could be effective; the rule was dropped.

High School

I later went to the highly selective Sydney Technical High School, but was influenced very little by the teachers. I tended to pursue what I liked, which included art, archaeology, geology, history, French and German, and poetry, irrespective of the examination syllabus. Chemistry fascinated me from the beginning, with an aesthetic rather than an intellectual appeal.

I adapted to the system and became academically "good" after my father became ill. My rank was second in the school in my 4th and 5th (final) year although only about 5% of the children even went to high school. By then I knew I had to depend entirely upon myself.

I was no favorite with most of the teachers. A question I cannot answer is why I behaved in my early education like a visitor from Mars. It was not that I did not understand the situation. My fellow students knew that the opinions of teachers counted in terms of marks and prizes and progress. I cared about these in theory. In practice, my confidence in my objectives and viewpoints was dominant. I did "my own thing", not what I was told was acceptable.

I was therefore not a good examinee; I did not give the expected answers. I saw subtleties in questions that the examiners had missed, a lesson I remembered for my own career as an examiner. I must have had some good examiners, otherwise I might had gone nowhere. One teacher, "Snowy" Turner, mathematics and my form patron, was so good that he made me look brilliant in mathematics, with first class honors in the double subject. Still, I knew I was no mathematician, unlike my friend Clive Davis, who later became Professor of Mathematics in Queensland.

The Headmaster did not like me because I was too independent and hated contact sports. (I liked tennis, although I was inhibited by an eye problem, and swimming, but not competitively.) He refused to make me a prefect, which would have been a normal assignment for my academic situation; he thought I was not public-spirited and "matey" enough. I could not have cared less. I decided to be school librarian, which I managed to achieve without difficulty. I was born an internal alien to the Australian macho tradition. I managed to escape sports on

Wednesday afternoons by an obvious administrative dodge, which gave me a lasting contempt for sloppy organization. I used to read in the Sydney Municipal Library instead.

A little story shows the extent to which my academic aims were misunderstood at school. The Deputy Head (English), when I told him I wanted to be a chemist, said, "Oh, you need Latin for that, and the school does not provide it." Consequently, I studied Latin by myself for 6 months, carrying a Latin grammar everywhere. I recall sitting on a headland at Neilson Park, studying it instead of watching the beautiful yachts on the harbor. I was very serious. Eventually another teacher mentioned by chance that the first one was confusing the terms chemistry and pharmacy as they are used in the British tradition. The only Latin I recall is *Caesar adsum jam forte, Brutus aderat* (to be read phonetically), but the effort to learn it probably helped me with Spanish and Italian many years later. The school chemistry syllabus contained little organic chemistry, although quite advanced inorganic chemistry was offered, including coordination complexes. I continued to educate myself.

Motivations

The driving force that led to my later successes was sheer necessity, but fueled and directed by emotions generated God alone knows how. On that score I record, I believe relevantly, that one of the major personal influences of my early youth was a Congregationalist minister, Harold King. He was one of the best people I have ever known, sensitive, unselfish, with a sense of humor, very strong willed, and unflinching in the pursuit of his literal interpretation of the doctrines of Christ, which I did not share but understood. He was no weakling. He had been in the First Australian Infantry Forces at Gallipoli and in France, but in the Second World War was threatened with imprisonment as a pacifist.

Deep personal interaction with him generated in me a puritanical attitude of mind, a feeling that I had to justify my actions and conclusions to myself on the basis of clear principles (which had to be sought) rather than to care about what was popular or acceptable or received public recognition. I still quote Scripture from that font of superb English and poetry and ethics, the Authorized Version of the Bible, although I am a "failed" Christian: "And though I have the gift of prophecy, and understand all mysteries and all knowledge, and though I have all faith so that I can move mountains, and have not charity, I am nothing." Modern English has no real equivalent of *caritas*, which is a pity. This passage is a superb statement of our present social–scientific dilemma. When I quote it now, most of my students have no idea what I am talking about.

The Congregationalists and their heretical ilk achieved the Industrial Revolution, so their social attitudes must be relevant at least to the entrepreneurial applications of science. My father's drive for autonomy was also very influential in forming my personality.

Why Chemistry?

There must be a strong temperamental factor in my choice of a career, based on some deep personal emotional need to understand what lies behind appearances. Revelations of such understandings have led me to ecstatic "highs" that I am sure no drug could produce. Two examples mentioned later were the first metal–ammonia reduction experiment (described in the chapter called Research Set 2) and the first doodles for the polyketide hypothesis (described in the chapter called Research Set 3).

But why my particular kind of chemistry? I recall as a child being intrigued by the pleasant smells of the leaves of eucalypts and wondering what makes the leaves of different trees so different. The bright and strange Australian flowers attracted me. My interest was not so much botanical as puzzling over why this bottlebrush was yellow and another one was red. What was the red-brown gum on gum trees, and why did they make it? The unique Australian blackboys (*Xanthorrhoea*) particularly attracted me with their abundant aromatic yellow or red resins, and many years later I investigated them chemically. I was never interested in specific plants, only in the connecting generalities. Odors attracted me. My first original experiments at the age of about 12, financed by my aunt's legacy, consisted of steam-distilling leaves and resins. One result, the biologically significant identification of lemon-smelling (+)-limonene in a chemical form of *Araucaria bidwilli*, was later published.

Beautiful crystals always aesthetically appealed to me, which is one instinctive reason why I have usually avoided "messy" technical areas of great biological importance. Robinson once told me that his first interest in chemistry developed from the habit of Yorkshire school children of putting bluebells into ants' nests to turn them pink.

I have never been emotionally attracted by exactitude of detail, but rather by the broad sweep of ideas collected around philosophically defined examples that can be tested experimentally. I have never been a

Under a gum tree in 1987. The attractive smell of the leaves puzzled me as a child and perhaps contributed to the allure of natural-products chemistry later in life.

scientific "stamp collector", although it takes all types to make the world and I have greatly benefited by the "collections" of others, as I shall describe in the chapter called Research Set 3.

University Influences: Sydney, Oxford, and Cambridge

Sydney, 1933–1938

Research capability depends on two factors: training in appropriate practical techniques and ideas, and development of an attitude of mind that leads to innovation. I was fortunate in experiencing circumstances that would promote such an attitude.

I went to Sydney University in 1933 (at the nadir of the Depression) with no scholarship, despite ranking third in the State of New South Wales (about 3000 students) in chemistry. Rita Harradence (later Lady Cornforth), who came first, was awarded the only scholarship for chemistry. Although there were no other scholarships available to me, I was awarded an Exhibition grant, which meant that I paid no fees.

Thus, responsibility for my own livelihood was thrust upon me from an early age (17). Because of my father's illness, I could expect little help from my family, apart from a room. To remain at the university with no entrance scholarship, I had to win the only one available at the end of my first year, the Levey Scholarship (50 pounds) for Chemistry and Physics. I achieved this goal by ranking first in both subjects, with 100% in chemistry, which was unprecedented so far as anybody knew. To make any progress toward my inner objectives I had not only to be good, but to be the best. Thus I was forced to acquire autonomy and independence, work attitudes that I carried through life despite an inherent dreaminess and laziness and boredom with detail.

My desire to learn, which had grown even more acute, prompted me to earn my own way somehow. I cleaned bricks and had other menial jobs, as well as coaching failing physics students. My total success rate as a tutor proved a good prelude to teaching. I recently met one of my erstwhile students, who said, "I have been waiting for 50 years to shake your hand; you saved my career." He had become Head of the Australian Commonwealth Department of Primary Industry; if I had not pushed him through he would have had to abandon his degree in agriculture and his career. This aspect of tutoring is now satisfying to the ego; then it earned me 5 shillings per hour.

I recall standing in the Vice-Chancellor's Quadrangle on my first day at Sydney University (1933) and thinking, "Here am I now, surrounded by the knowledge of the ages and the people who understand it." I had an immense sense of privilege and of obligation. In the same quadrangle I later, as a professor, had disagreements with the Vice-Chancellor (historian Stephen Roberts) about inserting iron steps in the wall to permit escape from the upper-floor organic chemistry research laboratory in case of fire. He prevented this installation because it would spoil his view. Fortunately, our safety code was such that he has no incinerated students on his conscience.

I started research even at school. I was interested in geology. While a student in Geology I in my first year at the university I completed my earlier work (started with the microscope from my aunt's legacy) on the heavy minerals of the Triassic Hawkesbury sandstone on which most of Sydney stands. I was trying to see whether the ratios of constituents (magnetite, ilmenite, rutile, zircon, and monazite) could be used to distinguish the source of these strata from the conforming Narrabeen series. (Later research showed that they are different.) This work was later published and gained a student research prize.

The University of Sydney is the oldest in Australia (founded in 1851), had about 3000 students in 1933, and was then the only university in New South Wales (three times the area of Britain). It had an honorable tradition in organic chemistry. Robert Robinson, in his first Chair (1913–1915), was the first Professor of Organic Chemistry in Australia. He was followed by James Kenner and John Read. The professor in my time as student was the first Australian in the Chair, J. C. Earl. He was not a great but a competent chemist, much undervalued by history. He was apparently appointed because all of the brilliant British "pommies" preferred to go home.

Professor Earl kept the flag of research flying through a shortage of resources and in spite of a lack of communication with the international world of chemistry. The nearest university was then nearly 600 miles away in Melbourne. Even as late as 1937, scientific publication in international journals was slow. It took up to 6 weeks each way to reach the United Kingdom for communications, referees's reports, revisions, and so forth. Still, this delay was less than the time it took when my grandmother was a small girl: 3½ months by sailing ship from Ply-

In a favorite beach setting. The local Hawkesbury sandstone became one of my research topics while still at school.

mouth in the 1850s, she told me. Only with airmail (1938) and air traffic after the war was the sense of isolation somewhat overcome. It still took three expensive days, even in 1954. The challenge for Australian researchers then was to find what could be done under such circumstances that would be internationally meaningful. I shall return to this question.

My student courses in Sydney now seem to me to have been rather old-fashioned, even by the standards of the time. I was taught the 64 or so classical reactions, but first learned about the electronic theory of reactivity later in Oxford. Bonding theories had been awaiting people like Pauling, despite theoretical advances in physics since about 1914. Physicists did not speak to chemists in those days, particularly organic chemists, and vice versa. Stereochemistry, although taught in principle, was not deeply considered or applied to any complex cases except, very formally, with sugars. Although the volume of knowledge has probably doubled about every 10 years since then, it is now far more efficiently packaged theoretically and is less laborious to absorb.

University Teachers

I interacted at the university with some exceptionally creative chemists. What influenced me most was not the transmission of knowledge, which I could and did get through reading, but attitudes subconsciously absorbed, as I later consciously realized. One unpaid vacation job

(1936) involved preparing cuprous–arsine complexes for that neglected alcoholic genius George Burrows, and through it I acquired a feeling for coordination compounds that surfaced very much later. From the professor, J. C. Earl, I learned that creativity could make a practical impact under very unfavorable academic, financial, and social conditions, given appropriate strategic considerations.

Francis Lions, whom I admired greatly, partly because his temperament was totally different from mine, and because of his evident high ability (he was described to me later by Robinson as his best Ph.D. student), had not at that time lived up to scientific expectations. This shortcoming resulted partly from his attempts to exercise creativity although he did not have the practical means to do so in his Sydney situation. He was a Don Quixote, who, however, later slew his giants. By stimulating F. P. Dwyer he largely founded a new coordination chemistry with polydentate chelate complexes, the organic parts of which he synthesized with a theoretical anticipation of virtually all of the stereospecific and enantiomeric consequences later developed in practice. This was a field extended by Australians Ronald Nyholm, Alan Sargeson, and others.

Lions's seminal contribution is not broadly recognized because he and Dwyer habitually published in rather obscure Australian journals, partly because of geographic isolation, but also because they felt so *Australian.* He never, for example, took the traditional sabbatical to Europe until after he retired (when he came to us in Manchester). Dwyer had, even for me, an almost incomprehensible Australian accent.

Lions was a great teacher, capable of generating enthusiasm. He was an aggressive person, not in his self-interest, but on principles. He told me once that a drop of honey has more lubricating value than a barrel of vinegar, but he usually took little notice of that fact himself. He quarreled with almost everybody in sight, including Professor Earl, usually for good reasons. Not only intellectually brilliant (double university medalist, 1851 Exhibition Scholar), he was a considerable athlete, an Oxford Blue, vice president of the Sydney Sports Union, the Athletic Club, and Swimming Club, winning the Sydney–Goulburn bicycle race at the age of 37 and playing grade football while serving as Acting Professor. The broken nose he sustained in consequence was used in his lectures on chemical asymmetry.

As the only student-elected member of Sydney Senate for 10 years, among other insertions of thorns into the flesh of the administration, he saved from abolition (in the midst of a great public political storm) the organic chemistry Chair that I later occupied. He was part founder, with the Professor of Ancient History, of a jolly, widely-embracing dining club called the *Kabeiroi* (loosely, the fellow seekers). The group was based on Greek ideals that did not distinguish science from life, philosophy, history, art, or politics, or professors from undergraduates. Thinking of him, I once said, "To see the World upright, you have to stand on your head occasionally." He wrote some very amusing and pertinent "poems" about chemistry and chemists in the history of the University of Sydney, which I have published.[1]

Each year Lions used to give with great zest a spectacular lecture, involving cold light (luminol) that he illustrated by washing intimate underwear in the dark; timed reactions on which he used to take bets but always won; and other magic, finishing with a bubbling mixture of sulfuric acid and *para*-nitroacetanilide. After the fumes had cleared, he revealed a column of charcoal 2 m high. (This demonstration cleared the theater). Chemistry could obviously be fun for him. (Caroline in Jane Marcet's classic *Conversations on Chemistry* would have approved.)

Fellow Students

The years before the war (1935–1939) were unique in the quality of the students enrolled in Sydney science programs, not only in chemistry. Chemistry students included, among others, Alan Maccoll, John W. Cornforth (Nobel Laureate), Ron Nyholm (later Sir Ronald, president of the Chemical Society), and Rita Harradence (Lady Cornforth). They were a stimulating group with whom I used to go bush-walking in the unmapped Blue Mountains. Cornforth and Nyholm were a year behind me, perhaps fortunately for me, because I am not sure how I could have stood up against such competition.

With friends in Wentworth Falls, New South Wales, around 1937. From left: Dick Welsh, Edith Reilly, J. W. Cornforth (Kappa), and me.

Fellow University students at Wollongong, New South Wales, in 1938 and their subsequent fates. From right, front row: F. Boileau, public servant; and Bob Franki, high court judge; middle row: myself, Ernest Ritchie, Professor of Organic Chemistry, now deceased; A. Rosenfeld, cane farmer; and me; and standing in back, K. Sewell, medic (Australian Armed Forces Repatriation).

"Kappa" John W. Cornforth signed his initials in Greek (iota omega kappa), which he learned with Latin and French at school. He was and is a high achiever in chess, sports, languages, and literature. As an example of his wit, I quote one of his famous limericks (which he started to compose as an undergraduate during lectures he could not hear) that was written about me when I had wantonly offended him:

That Outpost of Empire, Australia
Produces some Curious Mammalia:
The Kangaroo Rat,
The Blood-sucking Bat,
And Arthur J. Birch, inter alia.

As a classical scholar, he would not have been accepted for chemistry in Manchester in my time. His high literary ability in the presentation of his work is very unusual, set against the accepted style of scientific papers, which is often characterized by flatness or by pretentiousness confused with profundity. There is a lot to be said in favor of starting science study with classics, like N. V. Sidgwick and Sir Cyril Norman Hinshelwood.

In at the Deep End: Research in Sydney

My introduction to research in Sydney was traumatic, but effective. I am grateful now for the confidence in me shown by J. C. Earl, my supervisor for an honors (4th year) and M.Sc. (5th year) degree, although at the time it seemed like neglect. He gave me independence to make strategic and tactical decisions leading to my own successes and failures. He put on my bench the "phellandrene" terpene fraction of *Eucalyptus dives* leaf oil (a commercial residue from the isolation of piperitone) and said, "I am sure you will find something of interest to do with that, preferably of commercial importance." Then he went off to Cambridge, England, on a year's leave. For that reason, four of the first five papers that I published carry my name alone. Some of the results are mentioned in the chapter called Research Set 1. Earl had an unusual scientific sense of proportion: as an example, at a time when combustion macroanalyses were the only useful form of information on molecular compositions and had to be performed laboriously, by hand, in a tropically hot laboratory, he often did these himself for his student Cornforth.

But, in any case, what research was suitable in a remote place, with no outside funding, lack of access to new chemicals and some common solvents in less than about 6 months, no highly trained students, no money for research support, and the most primitive equipment? This situation can still be found in many parts of the Third World; I encountered it many years later (1975–1987) in trying to help organize chemical research in southeast Asia. Even in 1952, when I became a professor in Sydney, the main organic chemistry research laboratory had one power outlet. The shortage did not matter because there was nothing to plug into it. Stirring was accomplished with a hot-air engine (Bunsen burner) attached to a shaft; I later gave the device to the physics department. Soft-glass round-bottomed flasks with corks (no ground glass) were, with water baths or sand baths, heated by gas.

When I was a student (1930s, but it was little different in the 1950s), Liebig condensers were available, but no fractionating columns, no hydrogenators (I made one); as physical methods, we had only an Abbé refractometer and a polarimeter. Stirring was done by a water motor, which often leaked or jammed. Distillation under reduced pressure required water pumps and produced a variable vacuum (12–20 mm, according to water demand), aggravated by leaking cork stoppers with collodion. This situation demanded inventiveness, skill, and indefatigability. I needed, for example, to learn glass blowing to make my own apparatus, and I still like to make glass animals. The only practical assets then were intelligence, imagination, skill, and dedication.

In 1937, my fourth honors year, the departmental research themes were natural products, limited examples of reaction mechanisms, and syntheses requiring few or readily available substances. Earl, for example, pursued some natural-product structures, as representing one of the few local advantages, and also specific mechanisms for which he could obtain the resources. Starting with work on nitroso compounds (on which he had begun as war research), he shed light on the very old problem of the mechanism of diazotization and was the first to prepare and transform aniline nitrite. In this work he and K. W. Mackney discovered the first mesoionic series, the sydnones, by using simple available chemicals. He also worked with John Cornforth on a novel triterpenoid saponin. Earl gave me a practical opening into that terpene area, the smells of which had attracted me as a schoolboy, at a level with which I could then cope (described in the chapter called Research Set 1).

Lions at that time carried out mostly synthetic reactions for producing heterocycles, for which opportunities could be based on what precursors could be easily obtained. He used to dun chemical manufacturers for free products like the new morpholine or acetonylacetone—a "sample" was a full Winchester bottle. He always worked on the grand scale; a Skraup reaction on a kilogram sample was fearsome to behold and the result, to his exhilaration, was mostly scraped in good yields from the floor and ceiling. When Earl was abroad, Lions supervised part of my work, unusually for him on a wood product, gmelinol, a lignan, which I shall discuss in the chapter called Research Set 1 as an example of structure determination. Lions also supervised Rita Harradence (Lady Cornforth) and Cornforth on a multiplicity of themes.

In looking recently at my M.Sc. thesis (1938, 25 pages) on α-phellandrene **1** and α-thujene **3**, I find that, for reasons I cannot recall, I only included work that I did without a supervisor. I guess I wanted to emphasize that the result was *mine*. (*See* page 60.)

An Australian in Oxford

Why England?

The hinge of fate on the door opening to my career was the award of a scholarship of the Royal Commission for the Exhibition of 1851 to Oxford in 1938. In those days no Ph.D. degrees were awarded in Australia. For a scientist there were only two paid professional avenues each year to Europe (or anywhere else): the "Rhodes" and the "1851". For the former I had no qualifications (on paper); I was not interested in sports, I had no public spirit, all I wanted to do was to excel in my work. It was ironic, therefore, that I much later found myself on an Australian Rhodes awarding committee chaired by Governor General Sir Zelman Cowen, with whom I did not agree. In the 1930s my knowledge of this very demanding situation partly motivated my dedicated work patterns as a student, and I seesawed or tied for top in my university year with Rita Harradence. In 1939 she and John Cornforth were jointly awarded an "1851". Rita and I shared the University Medal at our graduation.

In accepting a scholarship (there were two for Australia in any science subject) in 1938 I knew I was probably exiling myself (as the Cornforths did). Apart from rare university and CSIRO (Commonwealth Scientific and Industrial Research Organisation) jobs, there were at the time no opportunities in Australia for Ph.D. graduates. I

In 1938, on my way to Oxford to study as an 1851 Exhibition Scholar. Various "wars" lie ahead.

was also certain I was running into an inevitable European war. I went through the Mediterranean during the Munich crisis (September 1938), with fleets mobilizing and battered Spanish destroyers at Gibraltar. But I had no alternative if I wanted to follow my blind dedication to developing as a researcher.

England was still "home" even to most native-born Australians. It exerted a very strong emotional pull, and chemistry was not my only motivation toward Oxford. I had read my mother's school prize books from an early age and knew much about Britain and its history. I was fascinated by the idea of becoming part of the long historic British scientific and social tradition. I also began there in 1939 a major book collection containing virtually all of the original books of significance in the development of chemistry from Boscovich in 1763 up to about 1930. Assembling such a collection was not possible in Australia, I could ill afford it at Oxford, and I could not do so at all now.

Why Oxford?

As an ignorant Australian in 1938, I did not even know exactly where Oxford was; I thought it was a suburb of London. I received no help from the University of Sydney. The administrators even messed up the traditional "1851" free boat passage (first class). I had to pay my own way, steerage on E deck of the *P and O Strathmore*, in a cabin that was level with the water line and shared with three others, one a drunken lout. Only my frugal habits enabled me to find the required 47 pounds.

I chose Oxford rather than Imperial College London largely because of the reputation of Robert Robinson compared with that of Ian Heilbron. I was not ultimately disappointed with my choice, but not exactly for the reasons I had anticipated, which were solely related to my feeling of Robinson's greater creativity. Heilbron would have been a much better supervisor in the short term, but Robinson's neglect enabled me, in the long term, to "do my own thing". With Heilbron I might, like many of his students, have been influenced to continue in his research areas instead of needing painfully but finally profitably to find my own. Robinson's power and reputation assisted me later, but Heilbron was much more conscientious in immediately helping his graduates. I have recorded my reactions to Robinson elsewhere.[2] Briefly, they include an enormous respect for his scientific intuition and an equal exasperation with the way he often chose to waste it. Perhaps I should not say that, because Albert Eschenmoser has recorded that Robinson was asked which of his students most resembled him: "Cornforth? No, Cornforth is himself; Birch is the one who most resembles me." I can see what he meant, but am not sure that I take it entirely as the compliment I believe he intended.

In any case, I hardly saw him in Oxford, particularly during the war when, according to his secretary, he served on 37 committees. I could talk to him on Tuesday morning if I did so while he was tearing open envelopes, muttering to himself, and dropping them into the wastebasket. His secretary could not take shorthand and typed badly. I once asked him why he employed her: "She was in the Red Cross, you know!" A lesson I learned was the importance of a good secretary. He had no assistant directors of research. These functions were performed, if at all, by people like Cornforth and me without his knowledge or gratitude. His lectures, which he clearly never prepared, were superb for the best students. He would stop his discourse and begin to work out some creative new idea on the corner of the blackboard. Only a few really bright students could recognize his untidy innovative thinking patterns.

In 1938 I was apprehensive about Oxford. My only ideas about it, gleaned from novels, involved an aristocratic system with which I might not cope. I turned up in September, knowing nobody in Britain (I had some addresses of aunts and uncles on my father's side), with a colonial accent and brown shoes. I learned to ride a bicycle and set to work in the Dyson Perrins Laboratory. It was not as socially difficult as I had expected, but I largely wasted my college opportunities. I ate there regularly until I became so tired of the terrible food (like grey Brussels sprouts cooked to rags) that I took to buying chocolate biscuits at Woolworth's (8 pence a pound) and eating cold baked beans with a teaspoon. Our former Prime Minister, R. J. Hawke, who was there at the same time, recently recorded that he has never eaten sprouts since.

The subtleties of the English class system initially gave me problems. You could join it, or you could avoid it as I did. Being Australian with an outlandish accent automatically put me outside it, although some of my fellow Australians sounded more Oxford than Oxford after 3 months. The Oxford University Regiment took one look at my tie ("What school is that? Sydney Technical High School."), and that ended my chances as possible officer material. By not joining the system I missed some of the good things as well as the bad. However, I joined a number of Oxford societies, including the Oriental one. I met A. M. Bakir there, for example, and later helped with idiomatic English as he wrote his thesis on the status of slaves in ancient Egypt. From what I saw in 1981, as Professor and Professional Fellow of Lincoln College, Oxford has changed little and I very much. The result is probably good in both instances.

Oxford Research, Good and Bad Aspects

The university was fragmented into colleges and university laboratories, between which there was little interaction. As I recount in the chapter called Research Set 1, I was one of the few from the Dyson Perrins Laboratory (DP) to interact with the Physical Chemistry Laboratory (PCL). I needed to make use of their developing "black box" double-beam infrared spectroscopy for my organic chemistry research. Robinson and Hinshelwood, Head of the PCL, did not get along at all and were totally irreconcilable personalities. One consequence was that almost all of the important college fellowships (which doubled salaries and research student access) went to physical chemists, because Hinshelwood was good at playing the college game. Nevertheless, I was actually offered one at Jesus College (about 1944), but had no difficulty in refusing it and so escaping the unduly comfortable Oxford system.

The professors had a limited control of intake of students and academic staff, who were mostly "wished" on them by the colleges. It created a very difficult situation, apart from the very high quality of all students (such as Margaret Thatcher, a chemist in my time). In short, many virtues were subordinate to Oxford parochial politics. Derek Barton, unforgivably, was later not offered the Wayneflete Chair at Oxford for such donnish reasons. From my observations in Lincoln College (1981), such gentlemanly academic traditions (daggers used according to stately rules, together with the Cornford dictum that "For University Committees, nothing shall ever be done for the first time") still continue. Some university professors are discontented with the situation, particularly in science, where more overt rumbustious measures are called for in the modern world. The dreaming spires of Oxford are indeed conducive to dreams. In my time they showed little potential to lead further to reality in practical organic chemistry. The situation is, I believe, changing for the better.

The Dyson Perrins Laboratory

I rapidly discovered that I learned from fellow students rather than from the staff in the DP. There were then and later many interesting research students from all over the world, and even from Britain. Some like John Cornforth (Australia), Michael Dewar (Britain), Richard Martin (Switzerland), Leon Golberg (South Africa), Hal Openshaw (Britain), Jack Edward (Canada), John Barltrop (Britain), and many others were highly self-starting, intelligent, and creative. All of the 40 or so research students in the DP were very friendly and able, but many required creative scientific assistance that the system then did not provide.

Sir Robert Robinson in 1951. Jack Roberts recounts many stories about Robinson, Woodward, and other contemporaries in his volume The Right Place at the Right Time *in this series. (Photograph courtesy of J. D. Roberts.)*

Lack of Records in the DP. Robinson frequently did not make proper inquiries as to personal antecedents or organize entry procedures, and some very odd people turned up. One problem was that he had no proper system of correspondence or filing and wrote most letters in longhand with no copy. I never had a typed letter from him until he was President of the Royal Society (and frequently not then). The Executive Secretary, Sir David Martin, complained about having to deal with unknown correspondents on unknown topics initiated by Sir Robert, who had forgotten.

One student from the East turned up in the DP to be told there was no room for him. "Who admitted you?" "You, Sir Robert; here is your letter." A student came to say good-bye. "Congratulations on your degree; who was your supervisor?" "You, Sir Robert." Another student from the East arrived with a letter from his uncle, one of Robinson's former students, so he was accepted without inquiry. Robinson was then attempting a brazilin synthesis and needed to do a large-scale Dieckmann ring closure, so he told the new boy to clean up a large amount of sodium and powder it. He did so with a brush, a cake of soap, and hot water. The laboratory barely survived, and the student was fortunately unharmed, but he left hurriedly. Another one, an American, dropped all of the sodium from the storeroom into the Thames River and the drains in Broad Street on a Guy Fawkes Day during the war. The result was a major reaction with the Oxford police, and I was left with nothing to react in ammonia for 3 months. The same student failed repeatedly in efforts to make pseudostrychnine with 30% hydrogen peroxide. Finally, on about a kilogram scale, he evaporated the lot under reduced pressure on a steam bath. The resulting explosion, fortunately during the lunch hour, demolished much of the laboratory. Fragments of cast-iron steam bath shot through the windows like bullets. When last heard of, he was operating a gas station in Texas.

Unassessed Entrepreneurs. A Hungarian, George Muller, had an interesting entrepreneurial character, as attested by his chemical work on substitution reactions in 100% perchloric acid. He could trinitrate benzene in the cold, but explained that the yield was only about 30% because it exploded twice out of three times. (Robinson's comment to me on this statement was "Nonsense, he only trinitrated dinitrobenzene.") He worked in a glass box on the roof, a relic of World War I poison-gas research. There would be a bang, a cloud of black smoke, and a shower of glass. However, George was indestructible and, after dusting down, was as good as new. He tired of organic chemistry, made a lot of money from buying up damaged houses in London immediately after the war on a rising market with no capital, and became first a philosopher and then a Professor of Geology in Chile. He met the definition of a typical Hungarian: one who goes into a revolving door behind and comes out in front (and I do not intend that pejoratively). Michael Dewar, at one stage during the war, was doing something related to the explosive RDX. After losing several cupboard doors, he turned first to antimalarials and then to even safer theoretical themes.

Where can we find their like in these sanitized, certificated days? There was an air of alchemy and stimulating uncertainty in the Dyson Perrins Laboratory, where almost anything could happen, including revolutionary ideas crystallizing out of chaos.

No Laboratory Designer. Part of our practical problem was the poor design of the old DP (1915) and its standards of upkeep and organization. The experience taught me a lesson for future application. Even the new building (1942) was badly designed, without, for example, any power outlets near the too-small hoods. Robinson's attitude was "We don't often stir in fume cupboards, do we?" (He used test tubes and round-bottomed flasks and used to set fire to his evaporating ether rather than condense it; this process was safer, he alleged.)

Robinson, if no laboratory designer, had his human and attractive side, with an interest in bridge, chess, and music. I recall once when he was nearly blind, about 1960, taking him to luncheon in the Festival Hall. As we came out, behind a nubile lass in a miniskirt, he asked, "Do they have skirts like that in Australia?" His sight was somewhat selective.

A Boring Problem. My first 2 years in Oxford (1938–1939) were a time of somewhat agonized self-discovery. I had had some facile autonomous successes with terpenes in Sydney (described in the chapter called Research Set 1), but no experience with synthetic chemistry. Here I had to work on someone else's problem, without being convinced of its validity. Robinson asked me, at the request of E. Stenhagen (Lund), to synthesize long-chain fatty acids with a quaternary carbon adjacent to the carboxyl group. The theory was that the important toxic fatty acids of *Mycobacteria* (tuberculosis and leprosy) might have such a general structure. I doubted this possibility because of the reported ease of hydrolysis of the esters, a conclusion that did not assist my experimental motivation, and I never had any of the natural compounds to investigate. Later my admirably conscientious friend Nicholas Polgar, as the result of immensely patient fractionation work, dropping cigarette ash each day into a multitude of crystallization plates, showed them to be normal fatty acids with methyl branches.

My work[3] was eventually an excellent synthetic introduction to forming quaternary carbons for my next task in steroids (described in the chapter called Research Set 2), but at the time it was dismaying. I had to obtain a D.Phil. degree; I did, and it resulted in a publication, but I have no copy of my thesis. I destroyed it shortly after obtaining the degree because I regarded the work as unimaginative potboiling, even though one examiner, that kind man Heilbron, complimented me. Perhaps I was wrong, but that was my emotional reaction at the time.

What Now (1941)?

With the advent of the war (1939) and the completion of my degree (early 1941), what could I do? I had enlisted in the Home Guard (1941–1945) and reached the exalted rank of corporal, first as a machine gunner. Then, because I was too accurate in blasting targets and also inclined to argue with the British officers about ridiculous orders, I was made a medical orderly. If the Nazis had invaded, their route around London would have gone through Oxford. With 300 rounds of ammunition and a bayonet on a pole for defense, I am certain I would not have been here now.

Robinson had largely given up on reaction mechanisms by then, after taking a thrashing from Ingold on the benzidine rearrangement.

In my time he concentrated, often with patient Indian collaborators, on alkaloid structures and on steroid synthesis. It took him a long time to realize that success was much harder to achieve in the steroid field than it had been with his classical work on flavonoids and anthocyanins that involved only a few stages amenable to one bright idea.

My synthetic steroid work, aimed at assisting fighter pilots with cortical hormones, was initiated for reasons I shall discuss in the chapter called Research Set 2. In 1941–1945 it was largely based on attempts to utilize an available Robinsonian cyclopentenophenanthrene precursor **18** that had functional groups in the key 3,11,17-positions for cortical hormones. It and its precursors (like **19**) were largely aromatic in origin, and therefore lacked angular methyl groups. Robinson's usual optimistic view was that "somebody will find out what to do with it" in making biologically useful products. My attempts to bring this Robinson precursor to fruition in active hormone structures led me to devise novel ways to introduce angular methyl groups and to achieve partial reductions of aromatic systems. My resulting reactions turned out to be very valuable, with general applications in many other fields. One very major outcome was the metal–alcohol–ammonia process for partial hydrogenation of aromatic systems, now known as the Birch reduction (discussed in Research Set 2).

At the bench in the Dyson Perrins Laboratory at which the first Birch reduction was carried out. In color the 2,4-dinitrophenylhydrazone stains would show!

Collaborations. To achieve actual production goals in such a major synthetic field, on practical time scales, required teamwork that Robinson did not have the time or inclination to organize. Also, he did not adopt the later American approach, pioneered by W. E. Bachman and perfected by Robert B. Woodward, of polishing experimental reaction stages. After a few tries he thought of a different approach, a failing I have in common with him. Robinson's steroid total synthesis problem was in fact solved by a unique team built around Cornforth[4], who had the right systematic and organizational approach. Cornforth coped with Robinson partly through personality, partly through sheer ability, partly through selective deafness, and partly because he could beat him at chess. As mentioned in the chapter called Research Set 2, much later in Canberra, Ganugapati S. R. Subba Rao and I found out how to convert Robinson's precursors into classical sex hormones. However, we had no pharmaceutically applicable results.

Why Stay in Oxford? Under the English academic system, I had to depend on Robinson for research collaborators. He gave me only two for very short periods over 10 years: Renée Jaeger on his joint work, and Sailendra Mohan Mukherji (who was to be there for only about 6 months) on mine. Applications of my ideas were therefore slow. But for personal reasons I could not move from Oxford; I experienced great difficulties with the increasing paralysis of my mother, as I shall mention later.

Autonomy de Jure. In 1945 I was offered in Oxford, and accepted, one of the new ICI fellowships, along with people like Michael Dewar, Rex Richards, and Charles Coulson. I also had a similar offer at Imperial College, London, to work on my proposal on the chemistry of mesomeric carbanions, which I was then developing as a new topic (described in the chapter called Research Set 2).

About 1946 Joel Hildebrand offered me a post in Berkeley, which I had to refuse because of my mother. I was married in late October 1948. On the first of January 1949 I was in Cambridge as Smithson Fellow of the Royal Society, a prestigious and independent job, but with no specific research support. I owed the post to the good opinions and power of Sir Robert Robinson and Sir Norman Haworth, according to Sir David Martin.

Losses and Gains. Whether personal problems were deleterious to my work in the end is a matter of speculation. I might under different circumstances have pursued a conventional career with collaborators to work out conventional ideas, accompanied by standard teaching and organizational distractions. But this path would have inhibited some of the rather lonely creative developments torn out of me by my situation, which I believe did contribute in major ways to the chemical advances of the period.

Cambridge University, 1949–1952

I had had a degree of independence in Oxford, largely by neglect; in Cambridge I had it de jure together with a personal income that, although not luxurious, was not my previous total penury. But I was not a university staff member and I had no official research support apart from that generously provided by Sir Alexander R. Todd. He, without obligation, gave me Herchel Smith as a Ph.D. student. The result was fortunate for me in obtaining an excellent collaborator and for Herchel (I hope) as a later multimillionaire (described in the chapter called Research Set 2). Then Joe Quartey, from Ghana, joined my research. Todd also organized a grant from the Nuffield Foundation Oliver Bird Rheumatism Fund for steroid synthetic work, under my control, in a defined strategic area: the total synthesis of steroidal anti-inflammatory agents. I scientifically translated this project first as the synthesis of model steroid male sex hormones (described in the chapter called Research Set 2).

Because of urgent family problems my college (Trinity) affected me little. I liked the master, G. M. Trevelyan, with his bristling moustache and his view of history as social development, including science and technology. However, the Cambridge Chemical Laboratory introduced me to broader questions and responsibilities of organization than those posed by my own research needs. Although I had an outside Royal Society appointment, I was graciously told what a great honor my appointment as a member of Faculty was. This attitude puzzled me. Why should an obligation to attend boring and personally irrelevant meetings, wasting time from real work, be regarded as so significant? Shortly, in Sydney, I found out why.

Oxford versus Cambridge

The organizational contrast at the time between Oxford and Cambridge greatly impressed me. The research population in Cambridge was as cosmopolitan as that in Oxford but was less fascinating in a human sense, being much better chosen for purely technical research ability. Nevertheless, or because of this, I made many lifelong friends, including Lord Todd, who did not teach me chemistry but introduced me to strategic ways of looking at research and its organization. I did not and do not always agree with him, but thereafter my first thought on organization was "What would Todd think about that?" Inversely, I thought, "What would Robinson do?" and I immediately considered the opposite.

The equipment in Oxford in 1938 had been the alchemical material that I described for Sydney. Tim Jones told me that when he took over in 1955 he ordered the smashing of several tons of useless, soft, badly designed, cheap glassware that still remained. When I had arrived there in 1938 I obtained the most basic equipment only through the kindness of those older inhabitants Hal Openshaw and Leon Golberg, who had their own accumulations. Later, as a wartime fire watcher, I raided the storeroom at night. My 1947 photograph shows virtually all of the ground glass then in the laboratory. In Cambridge in 1949, by contrast, equipment was all readily available on justified request and was as up-to-date as possible, organized by an excellent technical staff under the laboratory manager, Ralph Gilson.

The excellent Cambridge technical organizational systems contrasted with those of Oxford where, by default, the analyst, Fred Hall, both kept the financial notebook and looked after the equipment. In his own words, "All research is a waste of money", so the results can be

In the Dyson Perrins Laboratory about 1947, having accumulated most of the ground glass equipment in the laboratory.

imagined. "Why do you want a rubber bung? Perkin [William Henry Perkin, Jr.] never used rubber bungs." He was an excellent analyst, carrying out macroanalyses on 25-mg samples, but that was all. I once asked him, "Are you certain this analysis is correct?" "You can only be certain of one thing, my boy, death." I was sad when he eventually committed suicide. I rather liked the miserable old so-and-so; at least he was honest.

The selected few who required his reluctant services often found my Oxford laboratory technician, "Burdie", reading comics behind the gas meters in the basement. The technicians who gathered down there customarily used the stacks of university chemistry books dating back to 1600 to provide resting places for their wet tea cups. Robinson wasn't concerned; "But those are all out of date!" I got the books transferred to Sherwood Taylor at the Old Ashmolean Collection.

Cambridge science was at the same time more liberated and more restricted than the corresponding work at Oxford. As a student or a research fellow in Cambridge I would probably not have been permitted to develop the Birch reduction because it had nothing to do with the main research topics. Robinson had tried to stop me: "You are being paid to synthesize sex hormones, not to mess about reducing things in liquid ammonia", but by the time he found out it was much too late. In Cambridge the "assistant director of research" organization led to very close supervision on a few centrally chosen strategic topics: "Which test tube have you shaken since I saw you last, and why?" This attitude is exaggerated here, but only slightly. It is significant that several of the main Cambridge problems were solved by outsiders: H. G. Khorana with nucleotide synthesis (he was there to degrade Kenner's peptides) and S. F. Macdonald with the puzzling succession of aphid pigments through his little pocket spectroscope, a method learned from Hans Fischer (he had escaped from Germany as the war broke out). Macdonald eventually could not stand this system and returned to Canada.

Robinson's comment on Cambridge, "They are not chemists there, just a lot of paperhangers", referred to the frequent use of paper chromatography by Todd and by Fred Sanger. The point, as I saw even then, was to use whatever techniques worked, not to have a preconceived idea of what is respectable organic chemistry. But Robinson was one of the first to try chromatography on cloth draped through Perkin's "spy window" into a downstairs DP laboratory, in attempted separations of anthocyanins. He failed because he did not take solvent evaporation into account.

I have been exercised ever since by how to marry valid individual initiatives with broadly perceived desirability of results and with necessary teamwork. I find that it is almost impossible not to incline toward either chaos or over-control. Although I temperamentally tend to the former, I have tried in practical assignments (as in helping to organize the Research School of Chemistry) to combine the best aspects of both the Oxford and Cambridge systems.

Research in Cambridge

Extending Steroids. My personal research in Cambridge, starting in 1949, independently pursued the synthesis of steroid hormones. My student Herchel Smith was busy extending my Oxford victory of 1949 (in making the first totally synthetic biologically active male-sex-hormone anabolic agent, 19-nortestosterone, as described in the chapter called Research Set 2). He had been making what was later to be called norethindrone into the female progestational hormone 19-norprogesterone when I received a letter from Carl Djerassi in early 1951. It informed me that Djerassi had made 19-norprogesterone by partial synthesis, using my reduction method (described in Research Set 2). I stopped the joint work, probably mistakenly, as it turned out.

Literature into Realities: Analyzing Natural-Product Structures. I was becoming rather bored with synthesis in 1949. The general natural-products atmosphere of Cambridge turned my thoughts toward natural-product structure determinations and to a new topic for me, biosyntheses. The pursuit of these goals required inspiration, but less perspiration than synthetic work. It looked like a sensible turn, given my lack of personal laboratory assistance.

This ideological transition was helped by Todd, who had first wanted me to look at the admittedly important area of phospholipid chemistry. I (probably mistakenly) could not arouse emotional enthusiasm for such a messy experimental subject. He then suggested that I look at the structure of a crystalline natural plant anthelmintic protokosin (of my later β-triketone type), which he had previously isolated.

I started then to investigate the general topic of natural products by examining the literature for experimental evidence bearing on undetermined structures of reported natural substances. Several key personal names appeared. A. R. Penfold, J. L. Simonsen, and Harold Raistrick were excellent experimentalists, but unimaginative deducers of structures. I already knew this from Robinson, who used to help them interpret their experimental findings (usually without putting his name on the papers; he was satisfied to be the deus ex machina).

This path led me to "paper chemistry" and rapid deduction of about a dozen correct natural-product structures (e.g., as described in the chapter called Research Set 1). In a few cases I was able to obtain natural materials to confirm experimentally my deductions by using novel experimental methods based chiefly on my predilection for reductions and base-catalyzed reactions. I was fortunate also to be able to collaborate with F. N. Lahey, then visiting Cambridge from Brisbane, on the structure of the *Eucalyptus* sesquiterpene aromadendrene **11**, R = CH_2), which we defined as the first of a structural C_{15} series containing a cyclopropane ring.

Biosynthetic Pathways. A much broader importance of this Cambridge-inspired structure work, it turned out later, involved new structural and biosynthetic thinking. Initially my interest centered around Todd's kosin (anthelmintic) series of natural methylated and isopentenylated β-triketones, to which protokosin belongs. I then defined this general structural–biogenetic class, a very simple set of which is the acylphloroglucinols (described in the chapter called Research Set 1). The structure work thus had a very unexpected input into two major biosynthetic ideas I initiated in 1951 (described in the chapter Research Set 3): polyketides and carbon methylation and isopentenylation.

Being a Professor

University of Sydney, 1952–1955

In 1951 I was approached in England by Vic Trikojus, one of my former Sydney teachers, as an envoy of the University of Sydney, to see whether I would be interested in taking the Sydney Chair of organic chemistry in 1952. He was frank about the problems. That Chair (Head of Department) had been vacant for several years as a result of the premature unhappy departure of Earl. Ray LeFevre had made attempts to abolish the organic chemistry Chair. This attempt having failed, the situation then erupted in the columns of the *Sydney Morning Herald* in support of the "pretender" to the Chair, Francis Lions, who, however, failed to obtain it. Some years later I was appointed.

A very basic difficulty then was the extreme competition for the limited available resources. Sydney had few highly trained research students (the Ph.D. was introduced only in 1946) and little research funding. In 1955 my total departmental grant from the university for all staff research support (including chemicals, equipment, and posts) was 1875 pounds Aus., corresponding at most to about $50,000 in 1995. There were no other Australian sources of funds.

As an example of the situation, in 1953 I counted only three items of equipment in my department worth more than 30 pounds Aus., two of which I had acquired myself with outside funds (ICI). While in England I had kept in contact with my fellow student Ernest Ritchie, then a lecturer in Sydney, with whom I had attended primary school and university. In considering the Chair in 1951, I saw this difficult position through his eyes on the spot.

Why Return to Australia: Any Advantages?

To move my paralyzed mother physically would have been difficult, but she died in October 1951. By this time Jessie and I had three children, including twins. (My ex-collaborator Renée Jaeger's comment was, "That so-and-so Birch, 200% yield as usual!") I regarded Australia as a much better place for the children than Britain. Also, I was homesick for sun and sand and waves after 14 years, and I have never refused a challenge. I decided, with my wife's concurrence, to accept the invitation.

World-Level Research in Sydney in the 1950s. The question facing me was the same one that had faced my predecessor Earl: What kind of world-level research can I contribute in an isolated place with only the most primitive equipment and virtually no funds, where it can take 6 months to obtain chemicals, and with only commencing (but bright) research students who mostly do not proceed to Ph.D. level? Clearly, my steroid synthetic work was largely excluded (although I did accomplish some significant work, as it turned out[5]).

What were the local advantages, if any? The basic answer, probably, was the necessity to use my brains in lieu of anything else. On the practical side, Ernest Ritchie had been deeply involved with the Phytochemical Survey of the Commonwealth Scientific and Industrial Research Organisation (CSIRO). It had been first set up during the war, chiefly to look for useful alkaloids. Ritchie and others (notably G. K. Hughes) had sought new natural products for interesting new structures and biological activities.

CSIRO Natural-Product Advantages. The organization had the great virtue of providing plant materials in bags with botanically certified labels. This was an advantage with plants that did not grow in the United States, the United Kingdom, and elsewhere, especially for me. I could hardly tell a cabbage from a palm tree. Such structure work is very suitable for research training with beginning students. I hoped that a number of small results could merge into something bigger. But what general aspect could link such single diverse results?

A General Program? I thought of biosynthesis in 1951, but with the qualification that I must avoid alkaloids, partly to prevent cutting across the existing work on alkaloid structures of my friends E. Ritchie, G. K. Hughes, F. N. Lahey, R. Price, C. C. J. Culvenor, R. G. Cooke, and others. Besides, the theoretical basis of that series appeared to have been laid, so far as the organic chemist could contribute, following the ideas of H. Winterstein and G. Trier in 1910 and later of R. Robinson and C. Schöpf. Although my assumption turned out to be more apparent than real, I saw little opportunity then of contributing something genuinely new.

I looked at lists of other chemical types of natural-product structures in the Australian-derived literature. My major clue was the structure of the New Guinea compound campnospermonol **58**, which had been investigated in Australia. Its possible biochemical origin led me in 1951 to the logical deduction of the major polyketide biosynthetic hypothesis and some very important related "protective" ones. (Details are given in the chapter called Research Set 3.)

Learning to be a Professor

I arrived in Sydney in April 1952 as a complete innocent in administration, with little teaching experience. I had been elevated from a research fellow with responsibility only for myself into the Head of a large Department with, among other matters, responsibility for teaching nearly 1000 students.

We had two chronically ill children, no money, and no car. Personal problems abounded, especially for my wife. In university affairs I was greatly helped by F. Lions, E. Ritchie, and G. K. Hughes to survive in spite of very complex university politics, at which I have hinted. The doors between the inorganic and organic chemistry areas in my time as a student were nailed up with battens, and all contact between them went through the street. This attitude continued in 1952. The staff was trying to preserve organic chemistry as a discernible independent discipline.

Later I realized that I may have reacted unduly to this "survival" syndrome at the cost of risking the academic unity of chemistry, which I felt instinctively and later promoted in Canberra. But at the time I had to learn rapidly on my feet as a partisan administrator. I found administration astonishingly like research in many ways. It has a standard grid of theory within the spaces of which, as in organic chemistry, you had to probe for pragmatic results. I concluded therefrom that chemistry is a good training for administrators. In physics and mathematics the theoretical grid is often too fine and in many aspects biology is too broad to inculcate attitudes pertinent to the conduct of human affairs. I was fortunate in Sydney to have David Craig as a sympathetic colleague. Our cooperation proved to be a precursor to our foundation of the ANU Research School 10 years later.

The university was given finances for a new chemistry building by the New South Wales government, mainly because the State Minister for Health, M. O'Sullivan (formerly the Minister for Trams), as a typical politician, did not know the difference (confused in English wording) between chemistry and pharmacy. The State government wished to promote pharmacy by establishing a degree program. I well recall imparting this difference to the astounded Minister with some glee at a party also attended by my friend Roland Thorp (pharmacology). The Chemistry School, with support politically committed, was reluctantly accepted by the government when the situation became clear. Pharmacy moved into the old Chemistry Department facility, where it still remains. In 1953 I received a Carnegie fellowship to examine the construction of laboratories abroad. This was my first trip to the United States, where Carl Djerassi kindly and efficiently organized a scientific tour for me. The Sydney chemistry building was erected after I left in 1955. The ideas gained in helping to design it assisted me to participate effectively in planning the new laboratories later in Manchester and in Canberra.

My Science in Sydney and Its Support

I wanted to test my biosynthetic hypotheses of 1951–1952. I had been able to correlate many mold metabolites structurally for the first time as polyketides (Research Set 3). From this correlation I concluded that microorganisms, rather than higher plants, were the practical instruments into which isotopically labeled precursors such as acetate and methionine could be incorporated for test purposes by the solely chemical means available to me. Relevant isotopes, notably ^{14}C, were then becoming available because of isotope developments during the war. However, virtually nothing had been done in fungal biosynthesis, except in a few lipid (fat–steroid) fields.

In 1951, before leaving England, I talked to Harold Raistrick, the world leader in the isolation of mold metabolites, and suggested collaboration in this area. He was ill, thinking of retirement, and very discouraged by his earlier failure (as he saw it) with penicillin, a story not usually told. He did, however, make some very helpful biological suggestions, which I acknowledge with pleasure. Moreover, he supplied me with a strain of *Penicillium griseofulvum* as the source of 6-methylsalicylic acid (**67**), an obvious although slightly modified polyketide (Research Set 3).

Testing Hypotheses. For practical purposes, I had to obtain collaborators trained in radiotracer work and in growing molds. Therefore, I required support outside the minuscule Sydney University funding for organic chemistry. The research grant of the entire Faculty of

Science for all purposes, even in 1955, was limited to 20,000 pounds Aus. Even with the then much lower costs, such as 700 pounds for a post-graduate scholarship (my salary, linked to teaching, was 2200 pounds a year), it was very little.

Accordingly, I wrote in 1952 to every organization I knew, asking for support. Todd, through the Nuffield Foundation, was generous. I was also supported from a most unexpected direction, and I recount a moral story. While in Cambridge one Saturday afternoon in 1951, I encountered a pleasant little American who said he was a medic, wandering in a lost sort of way about the deserted laboratory. I took pity on him and took him to tea at the Friar House, chatted egotistically about my ideas, and thought nothing further about it. When I wrote to the Rockefeller Foundation in 1952 I received a puzzling letter signed "E. Pomerat", a name that meant nothing to me. In effect it said, "We have given no funds to Australia since the Medical School debacle in Sydney in 1936, which upset us so much that we had removed the country from our list. Since, however, I know you personally, come to see me on the 65th floor of the Rockefeller Plaza, to discuss what you need." It turned out of course to be the little American, who then found a small black notebook, among many in his files, in which he had recorded our conversation. He smiled: "How much do you want?" It was not a lot. His grant was valid in connection with the medical interests of the Foundation at the time, because my ideas and work provided the background for a number of antibiotics, including the tetracyclines.

This dual key assistance permitted me to have Ralph Massy-Westropp trained by Bernard Ralph to grow molds and Clarrie Moye trained to measure ^{14}C-radioactivities. The first result, using methylsalicylic acid (**67**) from Raistrick's suggested mold, was published in 1955 (Research Set 3). It provided the critical biochemical support that converted my polyketide hypothesis into the major theory that now explains the biosynthetic skeletal origins of many thousands of natural products.

Advantages and Disadvantages of Being at the End of the Earth. What had first appeared to be a grave disadvantage in going to Sydney, the inability to continue total synthesis, turned out in the end to be a scientific asset. It is amazing what can result if you have nothing to use but your brains. As also scientifically exemplified in my organometallic work (Research Set 4), a new point of view imported from an "alien" field can lead to new understandings, not just to amplifications of old ones.

Although not entirely experimentally frustrated in Sydney, I was still unhappy with my practical facilities. I did not have even a UV spectrometer; some of my UV spectra were done on a little-used CSIRO instrument behind the Sydney abattoirs, and some IR spectra were provided by my old friend in Oxford, analyst A. Strauss (now alas dead), both free by grace-and-favor. One difficulty arose in consequence, which I shall recount later in connection with zierone (16).

Moving On

My dissatisfaction was known, and in 1955 I had two offers. One was from Sir Mark Oliphant, to take a foundation chemistry Chair at the School of Physical Sciences in the new Australian National University, despite some opposition within that institution. It is an interesting speculation what might have happened within the ANU if I had accepted: the school would have been what its name indicates, not a school of physics, as it is now. The Research School of Chemistry would not exist. Such are the hinges of fate. However, I refused in favor of an offer at Manchester to replace E. R. H. Jones, who was elevated to Oxford.

My departure from Sydney in late 1955 was accompanied by Australian newspaper headlines like "Beggars in Mortarboards. Why the Professor Resigned". The departure of Craig and Nyholm soon afterwards reinforced my message. As told to me in 1960 by Prime Minister R. G. Menzies, this exodus was a final stimulus for the Murray Commission of 1957, which revolutionized the financing and organization of Australian universities under Commonwealth auspices over the "dead bodies" of the States. I probably made my best contribution to the Australian university system by then publicly quitting it, although the undesirable 1990–1991 tertiary education takeover actions of the Commonwealth government justify the suspicions of the States in 1954.

University of Manchester (1956–1967)

I owe my opportunity in Manchester ultimately to Lord Stopford, the greatest vice chancellor I have known. A medic by background, he had high intellect and academic judgment, a great sense of proportion, and an ability to interact with people. He met my wife once; about 6 months later at a large gathering without me, he said, "Hello, Mrs. Birch, how are you enjoying Manchester?" In his time Bernard Lovell acquired the Jodrell Bank radio telescope, and R. J. Williams got the first major computers.

I have had the good sense, despite offers (knowing my "level of incompetence" and boredom threshold) to refuse to become a vice chancellor. Stopford and his successor, Mansfield Cooper, continued the enlightened administrative tradition of Manchester, which went back to the 1850s to Sir Henry Roscoe, the Professor of Chemistry and simultaneously a member of Parliament, and to Sir Henry Meiers. For a century Manchester was the greatest scientific university in Britain. The unspoken motto was "people first, equipment next, buildings last". H. Schorlemmer, a friend of Friedrich Engels and the first Professor of Organic Chemistry in Britain, taught there. Its faculty has included some of the most eminent chemists and biochemists (and also Nobel Laureate physicists like Rutherford and Blackett), including a number of Nobel laureates. On its list are E. Frankland, R. Harden, Alex Todd, Robert Robinson, Ian Heilbron, Henry Roscoe, W. H. Perkin, A. Lapworth, E. Evans, M. Polyani, and many others. This is not to mention Manchester associations with Dalton (whose bones I tried to save later, unsuccessfully; they are now playing football over his head).

Why Was I Appointed?

Manchester habitually appointed to Chairs young people on the way up (many in their twenties) rather than distinguished ones on the way down, as frequently happened with Oxbridge (the British abbreviation of Oxford and Cambridge Universities, to distinguish these ancient universities from the "red brick" institutions like Manchester formed in the 20th century). I pointed out this contrast unkindly to Todd, as a new incumbent in Cambridge (how to gain friends and influence people). He had defined Manchester as the place people "come from". I was rather old, in my thirties. Still, an illustrious tradition based on industry funds and equipment is difficult to continue when science can be pursued in more pleasant physical surroundings, where funds are now available. Manchester University (then "black velvet") now boasts a new cleanliness, and turns out to be pink, except for black tiles mistakenly replacing ones broken during the War by anti-aircraft fire.

I liked Manchester; it was easier to get out of than, say, London (to the Peak District, to the Lake District, or to North Wales, the loneliness of which reminded me of Australia). Also, I found the kindly and relaxed North Country people and accent congenial. In 1955 I regarded the offer of the Chair not only as a great work opportunity, but as a great honor to become incorporated into an illustrious tradition.

In a difficult board interview, having flown me 12,000 miles in 3 days at great expense, the committee had some problems about what important questions to ask me. Stopford eventually saved the situation by saying, "Well, Professor Birch, have you any interests apart from your work?" "I have four children, if that answers your question!" This response broke up the solemnity and appealed to his rather specific sense of humor.

In 1956, when I accepted the position of Professor at the University of Manchester.

My Establishment-supported rival was Alan Johnson, one of my favorite people, a close friend from Cambridge days (we worked in the same laboratory), and a very good scientist. He and I continued to be excellent friends thereafter until his sad premature death, but the establishment continued to be displeased. Todd recently told me that this was the only time he was really annoyed with me. He claimed that he had sent me to Australia to reform chemistry there. I had no particular support for the appointment to Manchester even from Robinson, who for some typically eccentric reason favored R. W. Bradley, the color chemist from Leeds. My strong protagonist was Tim Jones. Typically, Manchester and its vice chancellor made their own uninhibited choice.

Manchester Advantages

I could do in Manchester what I had wanted to do in Sydney. In particular, Jones had established a first-class laboratory for growing microorganisms. I was able to set up a tracer laboratory with Herchel Smith, who had agreed in 1956, as a condition of appointment, to learn tracer techniques to help test my biosynthetic hypotheses, which he did with great efficiency.

Although some of the specific steroid synthesis problems I had originally envisaged in Oxford had been pursued in the meantime by Djerassi and others, I was also able to re-recruit Herchel for joint synthetic research. He had worked with Robinson on steroids in Oxford after graduating with me. We later parted ways on the steroid work, as noted in Research Set 2, but the synthetic capabilities provided by the Manchester laboratory and its philosophy led him independently to the production of norgestrel, now probably the most used type of contraceptive pill constituent.

The Manchester laboratory was very well equipped for its time with IR and UV spectrometers. Also, we were able to persuade the University Grants Commission that a new building was required to replace the "interesting" ones built in 1874 and 1908. (The first was a copy of Justus von Liebig's laboratory in Heidelberg and the second was a precursor of the Dyson Perrins Laboratory in Oxford. They were good buildings, but structurally inflexible for changing needs.)

The importance of flexibility for unforeseeable developments was a lesson I then learned for later application at the ANU. The resulting new Manchester building was accompanied by new equipment. Its construction gave me experience in dealing with architects, which also proved valuable for later application. I pressed to obtain my Australian friend R. S. Nyholm (University College London) and Duncan Davies (ICI) as assessors in lieu of R. D. Haworth, a classical chemist who worked in round-bottom flasks and test tubes. In this way we became, temporarily, the best equipped chemistry department in the United Kingdom, with the best NMR and mass spectrometers available at the time. We also played a major part in designing the inlet system for the Associated Electrical Industries (AEI) MS9 mass spectrometer, to make it suitable for the now-classical organic usage (Research Set 1).

Continuing Research. I was very happy then (1956–1965) to be able to provide experimental support for many aspects of my published biosynthetic hypotheses, which had become very diverse (Research Set 3). I was pleased with the almost complete success rate of my predictions. This consistency was the result of applying the rigid criteria I had set up: structural coincidences, acceptable chemical mechanisms, biochemical probabilities, and personal instinct. Fortunately, much of the accumulating biochemical information of the time concerned primary metabolism routes that I could mechanistically extrapolate, as novel ideas, to new types of natural structures.

Novel Interests. Between 1956 and 1967 I pursued a number of interests that were complexly interrelated and mutually contributory. These areas included natural-product biosynthesis, new-product isolation and structure determination by novel physical and chemical methods (Research Sets 1 and 3), and the total synthesis of natural-product structures (including the biogenetic polyketide class) by novel regio- and stereospecific organic methods based on the availability of specifically substituted dihydrobenzenes from Birch reductions (Research Set 2). A more remote (but fully related) novel topic comprised organometallic reactions employing superimposed enzymelike reaction capabilities. We deliberately tried to rival in the laboratory some of the synthetic capabilities of enzymes (Research Set 4).

Contributions to Personal Reputation or to Science?

This list of fascinating programs was too long for my personal reputation, at least in the short term, and for my support facilities. I had no sooner been recognized as an expert in a given area (for example, by being asked to organize the Insect Chemistry Symposium at the 1971 IUPAC (International Union of Pure and Applied Chemistry) meeting in Boston or to be a plenary lecturer at the 2nd European Biochemical Societies' Congress in Vienna in 1965) than I had about ceased to qualify. I recall a discussion with Todd in Cambridge, who asserted that, "No laboratory can hope to make an impact unless it confines itself to a few fields." I understand his reasoning, but fields do fertilize each other, and I am temperamentally unable to avoid boredom under such circumstances.

H. G. Khorana solved the nucleotide synthesis problem while I was in Cambridge, because he was outside the Todd group there. He achieved this synthesis not in spite of the fact that he had no initial background in it, but because he had none. He had been imported to Cambridge by George Kenner to degrade the peptide hormone ACTH. However, he recognized the potentialities of the carbodiimides he had been using for peptide degradations within the Cambridge nucleotide synthesis field. As Todd said, "His work grew out of the atmosphere of the laboratory in Cambridge"; it is a major example of cross-fertilization between apparently unrelated fields. But attitudes, however acquired, matter; "Luck favors only those with an attitude of mind prepared to perceive it" (loosely Pasteur, who probably stole it from Goethe).

Not a Member of the "Club". While I was pursuing my widespread interests, several general features of the organization of scientific research became apparent to me as an indication of human limitations. For instance, I was never regarded as one of the club by "real biochemists" (or later "inorganic chemists"), despite my contributions to biosynthesis and matters like the genetics of flower color in *Dahlia* (Research Set 3).

Such pointed neglect had two bases, I believe. One objection is that I never used isolated enzymes; the other is that I published in "chemical" journals. In a joint Chemical Society–Biochemical Society symposium on biosynthesis, I was listed by the latter only as a member of the former. The situation did not last long, as I consequently resigned from the biochemists. This is an indication of the very narrow reductionist approach at the time to broad interdisciplinary scientific problems. Publications in chemical journals or biochemical journals (or inorganic or organometallic ones) seem to be mutually ignored by narrow protagonists of techniques and ideas, despite the "unity" of science.

Industry and Academic Research. Industrial connections assisted a number of practical aspects of my own work involving biological activities (Research Sets 2 and 3). They provided access to instrumentation, especially NMR spectroscopy about 1960 (in Roche, Basle, Switzerland), when this technique was only beginning to develop in universities. I also obtained some financial research support. I am grateful to people like Frank Rose (ICI), Leon Golberg (Bengers-Fisons, Holmes Chapel), Otto Isler and Bruno Vaterlaus (Roche), Carl Djerassi and George Rosenkranz (Syntex, Mexico, Palo Alto), and Ian Morton (Unilever, Colworth, Port Sunlight, Vlaardingen) for their insights and the assistance they provided. I applied their contributions later, both in teaching and in policy considerations.

A Complex of Universities within Britain, 1955–1967

Manchester's proximity to other universities provided a very pleasant contrast with Australia in the 1950s. Scientific progress depends on the evaluation of ideas by personal interactions and comparisons of standards, as well as through publications. Manchester is the center of the largest population concentration in Britain, with a number of first-rate universities over a radius of 50 miles or so. It is in fact "The Victoria University", which also once comprised Liverpool and Sheffield. A second university was being set up in Sydney in the late 1950s; Melbourne was some 600 miles away by train or DC-3. Armidale (New South Wales) in my time was a college affiliated with Sydney; I used to visit it as a professor (400 miles) to give a concentrated lecture course in 1953–1954.

A Manchester Administrator. In Manchester I dodged university administration, as distinct from departmental activities. Some years ago, when I received an honorary D.Sc. degree there, some astonishment was expressed concerning my subsequent political career (e.g., as Chair of the Independent Inquiry into the Commonwealth Scientific and Industrial Research Organisation). I learned the necessities of administration the hard way, as a member for a time of the Manchester Standing Committee of Senate, which did much of the university detailed business. I flexed my embryo Machiavellian muscles within organic chemistry by stealthily setting up and chairing the first Department of Biological Chemistry (now Biochemistry) in Manchester, with the assistance of G. R. Barker. I also set up an overall Board of Studies in Biochemistry (now with Molecular Biology), which correlates all biochemistry research and teaching in the university: faculty of science, medicine, and University of Manchester Institute of Science and Technology (UMIST). But that was another story on my learning curve.

Canberra Reconnections. In 1960 I attended the 1st Natural Products IUPAC Symposium in Australia, part of it in Canberra. I met some of the Australian National University administrators there, notably my old friend from Oxford, Sir Hugh Ennor, and also Prime Minister Menzies. This meeting led to discussions about the formation of a new Research School of Chemistry at ANU. Starting in 1962, the project involved me in fatiguing trips to Australia and in the burden of working out so much that was entirely new (1965–1967) as a joint venture with David Craig and our appointed manager, John Harper. It is not surprising that my research in Manchester suffered, although this decline was mainly in publication rather than performance by my collaborators.

We had persuaded the ANU to finance some scholarships and postdoctoral fellowships in the United Kingdom starting in 1965. With this support we could return in July 1967 with the setup of a new school with integrated work teams.

Research School "Administration". From left, David Craig, storeman Michael Griffiths, technician Brian Fenning, and I in Bramley's Bar Social Club in Canberra.

Completion of Manchester Research

My Manchester work had in the meantime (1960–1970) taken some entirely new directions, but I needed to make some hard choices. Under the university system I was required, as Head of the department, to ensure that my deserving staff (in various work areas) received appropriate research support. This support effectively restricted the number and quality of "bodies" available to me personally.

By about 1966 I had confirmed all of my many major biosynthetic hypotheses and had developed these in novel ways toward biosynthetic structure determinations, the partial synthesis of new antibiotics, and applications in phylogeny and genetics (Research Set 3). This field was becoming rather boring as ideas were confirmed. It only reawakened my interest later in Canberra with the availability of new NMR techniques.

I was faced in Manchester (from 1956) with some hard choices for personal research in a still rather restricted climate. For example, with very few postdoctoral fellows, research topics had to be tailored to the training of honors, masters, and doctoral candidates, rather than primarily to contribute to the chemical discipline. Several possible developments of previous work seemed feasible. One choice was the elaboration in the laboratory of my hypothetical biochemical β-polyketide chains, for the total synthesis of polyketides. We carried out the first deliberate synthesis resulting in dihydropinosylvin.[6] But having demonstrated that the approach works, we abandoned it for more subtle total synthetic procedures. It has since been adopted by others who have developed new ways to make our projected starting polyketo esters.

Another research direction (new for us) on which I had speculated was approaching biochemically the fundamental question of the nature of biochemical intermediates in polyketide biosynthesis. But this approach would have involved "real enzyme" biochemistry. After experimental tries, even though I was at the time (early 1960s) a de facto Professor of Biochemistry, I concluded that I am not a biochemical "cook".

The more personally attractive alternative was to turn back to synthetic work and toward "real" organic chemistry (which turned out eventually to be organometallic). This I did (Research Sets 2 and 4).

Steroids: General Methods Based on Birch-Reduction Products. After about 1962 I temporarily abandoned my direct experimental approaches to steroid synthesis in favor of Herchel Smith's personal effort (Research Set 2). However, I was still interested in planning new general stereospecific synthetic processes, particularly those radiating from the uniquely structured dihydrobenzenes obtainable from Birch reductions. A unique range of such novel synthetic precursors, containing unconjugated and conjugated cyclohexadienes, have specifically situated substituents like OMe (e.g., **36**), alkyl, and CO_2R (e.g., **118**) as reactive "handles" on one or both specifically positioned double bonds.[7]

Diels–Alder and Alder–Rickert Reactions on Birch-Reduction Dienes. I was aware of the regio- and stereospecificities of the Diels–Alder (diene) reaction, toward substituents on both the diene and the dienophile, from my commissioned literature summary for the Annual Reports to the Chemical Society in 1950.[8] The structural specificities of this reaction could not only be used to join two large fragments readily, but could, among other potentials, directly and uniquely lead to compounds with synthetically difficult aromatic patterns characteristic of the biosynthetic polyketide class of mold products (e.g., Research Set 2).

Synthetic Diels–Alder yields from dienes were significantly improved by the accidental observation of N. A. J. Rogers that the initial unconjugated 1-methoxycyclohexa-1,4-dienes from Birch reductions can be catalytically conjugated in situ (e.g., to **37**) by Diels–Alder 1,3-dienophiles. This process makes possible their direct use, as isolated from reductions without preliminary conjugation, and ensures their complete conversion in one step by re-equilibration of 1,4- into 1,3-dienes. Uses of such processes are discussed in Research Set 2.

Another very important novel synthetic outcome radiated from a casual observation in the course of our Diels–Alder work, under Research Set 2. An adduct formed with a bridgehead OMe was found to undergo acid-catalyzed fission of that one of the two newly formed C–C bonds that terminates at that position (e.g., **49**, *predictably* in hindsight). The overall synthetic result is to form a sterically controlled unbridged product with only one new C–C bond instead of the classically inevitable *two* new D–A bonds. The predictable stereospecificity from the original side-by-side (suprafacial) D–A addition can thus provide an important novel approach to the stereospecific synthesis from cyclohexadienes of some important natural products not containing bridged rings.

1,4- and 1,3-Dienes as Unforeseen Synthetic Precursors. One major unexpected outcome of this Diels–Alder work resulted from an attempt to prepare, as a precursor, a pure 1-methoxycyclohexa-1,3-diene **37**), instead of the catalytic equilibrium mixture (about 75:25) of both the 1,3- **37** and 1,4-dienes **36** discussed in Research Set 2. This approach followed an observation of my fellow countryman R. Pettit in the United States that unconjugated cyclohexa-1,4-diene, which is a primary Birch-reduction product from benzene, yields the $Fe(CO)_3$ complex of the conjugated cyclohexa-1,3-diene, from which the unsubstituted cyclohexa-1,3-diene can be released pure. We tried the 1-OMe series, and the result led to another intellectual "creation center" from which ideas and results radiated almost explosively (Research Set 4). The organometallic derivatives we thus investigated behaved experimentally like respectable organic compounds. I rapidly realized that their special reactivities, related to their symmetric and steric properties, could lead to a sort of "inorganic enzyme" synthetic chemistry. I believe that the rapid blooming of such ideas in the organometallic area marked a conjunction of my initial stereospecific organic synthetic work with my increasing knowledge, derived from my biosyntheses, of how enzymes must work.

My nascent "inorganic enzyme" ideas were further promoted by my work on a soluble hydrogenation catalyst. Having defined in Manchester, so far as was then possible, the mechanisms of the metal–ammonia electron-addition reductions (hydrogen additions) (Research Set 2), I turned to a complementary interest in methods of pure molecular hydrogen addition as uniquely exemplified by soluble hydrogenation catalysts.

Assisted by a very able collaborator, I set up a major review[9] to correlate published methods and results, experimentally and theoretically, and to assess critically the practical selectivities for laboratory workers intent on specific synthetic applications. This review took about 4 years to complete, and it certainly inhibited other work. Creative reviewing of this sort is grossly underestimated in the reward systems of science. A method often profitable in discerning something new in principle is to look critically and broadly from the outside, as we did, rather than to carry out trivial and often ill-directed new experimental work.

In conjunction with this review, we carried out the first extensive synthetic work using hydrogenations with the soluble chlorotris(triphenylphosphine)rhodium (Wilkinson's catalyst) with complex polyfunctional molecules, many of them natural products. The catalyst behaves almost like a hand in placing two hydrogen (or deuterium) atoms as a cis- unit. My aim, as a natural-product chemist, was to define the regio- and stereospecificities of addition within more complex molecules than those commonly dealt with by classical organometallic chemists and to obtain unique new products (Research Set 4). At a Gordon Conference in 1966 I suggested the obvious idea of asymmetric hydrogenation, as was recalled by E. J. Corey in a seminar I gave at Harvard (1987). Unfortunately for me, I was not then enough of a slave driver to compel the student involved to do the "donkey work" in resolving phosphines; I only *suggested.* We also reviewed in print the whole basis of organometallic chemistry as applied to organic synthesis. Establishing a new school in Canberra from 1965 inhibited my personal research, although it was partly offset by having no undergraduate teaching. I had many obligations outside the School and the university.

Research, 1965–1974. My biosynthesis area was winding down from 1967, climaxing with work on phomazarin with Tom Simpson (Research Set 3), who has since gone on to new heights in the general area. Some total-synthetic aspects were continued, based on Birch-reduction dienes, notably the development of Diels–Alder and Alder–Rickert reactions (Research Set 2) leading to my gradually increased appreciation of the principle I later styled "lateral control" of product structure in laboratory synthesis. This concept was principally developed experimentally later through transition metal complexation.

I continued to use the Birch-reduction products of substituted aromatics (Research Set 2) as experimental vehicles for much of my synthetic work of all types, because of the very large variety of pure compounds available with definitely positioned substituents and unsaturation, including unique structures such as regiospecific enol–ethers in unconjugated cyclohexadienes. And we knew how to make them.

From about 1962, but mainly 1967–1985, I undertook a broad case study in organic synthesis using organometallic compounds. It was a very laborious and systematic study, attacked from the viewpoint of the typical organic chemist, on the principle of the need of at least one functionally multisubstituted metal series. In my chosen case this was experimentally based on $Fe(CO)_3$ and cyclohexadienes (for reasons and details, *see* Research Set 4).

We obviously could not neglect known organometallic theory and practice, but our new emphasis was to examine aspects not normally covered by organometallic chemists. At the time they were concentrating on the transition metal atom and the nature of its bonding with simple, often unsubstituted, organic substrates. We were closely interested in the effects of complexation, not only on the reactivity of the combined organic portion but on functional substituents, as outlined in Research Set 4.

Unfortunately for me personally, apart from a few examples like gabaculine and isotopically labeled shikimic acid (Research Set 4), lack of practical support in my retirement has meant leaving exploitation to others, notably to my very able former collaborators A. J. Pearson and Richard Stephenson.

For 25 years I could not read much of the literature (I depended on collaborators for that), so perforce I was constrained to try to stay so far ahead of it that it did not matter. Of course, I received a few nasty shocks. I had to focus my attention on a few strategic areas I selected, rather than follow my inclinations and dissipate my efforts in many interesting new directions.

Despite all of the later distractions I was never out of my laboratories for long, although not at my bench. I wanted to feel that I was directly involved in the excitement of scientific detail, not just an organizational entrepreneur. I always talked to my collaborators in their laboratory and not in my office. I actively contributed scientific detail to every paper on which my name appears. I rarely exceeded six collaborators at a time, usually less. During 10 years in Oxford I had only two collaborators, each for short periods, although this lack was not through my own volition.

I also took Ph.D. training seriously, as intended primarily for the benefit of the student and only indirectly for my scientific career. I believed that students should make their own mistakes and have an inalienable right to fail, but only one of my Ph.D. candidates ever did. Difficulties came from the rather diverse topics comprised in my students' theses, from among which they had to select their choice for concentrated pursuit as results accrued. Academic committees want an exact title beforehand and tend to accept only work related to that title. A meaningful title that appears to be specific is hard to devise for theses containing diverse topics selected to give a spectrum of experience and training. The usual narrow academic approach is the antithesis of genuinely novel research. It is only good for exploiting the known. One of my very best students (Rod Rickards, Fellow of the Australian Academy of Science) never bothered to submit a thesis at all. Qualifications are in my view the ability to perform, not pieces of paper, but I had problems with that point of view.

Research Set 1

Research Sets

Research scientists fall into three groups: those who pursue a topic throughout their career (usually derived from their first supervisor); those who change careers completely for a number of interesting reasons; and those who alternate among disparate topics. I belong to the last group. This choice was temperamental, but also springs from a conviction that study of one topic can favorably influence another. In my case, the symbiotic topics are natural-product structures–biosynthesis, structures–synthesis (and vice-versa), synthesis–organometallic chemistry, and (in almost full circle) biosynthesis–organometallic chemistry.

My Research Sets are organized here on a subject basis, and coincidentally partly on an historical basis. The difficulty with constructing a time sequence is that I continued to investigate many main areas throughout my career, but with time lapses and overlaps caused by resumptions on new grounds:

- Natural products: Sydney, 1937–1938; Cambridge, 1948–1952; Manchester, 1955–1967; Canberra, 1967–1975
- Synthesis and mechanism: Oxford, 1938–1949; Cambridge, 1949–1952; Canberra, 1967–1980
- Biosynthesis: Cambridge, 1951–1952; Sydney, 1952–1955; Manchester, 1955–1967; Canberra, 1967–1975
- Organometallic: Manchester, 1960–1967; Canberra, 1967–1985

The interludes between research sets are usually historical, largely relating to my general situation between a preceding and following set.

Natural Products: Chemistry and Structure Determination

Why Work on Natural Products?

Why was this a prime subject for Australia and for many developing countries even now? I acknowledge that I personally did not seek new compounds and discovered few of those on which I worked. I am greatly indebted to friends like Ern Ritchie, Bill Hughes, and Maurice Sutherland for unselfish gifts of substances with no strings attached. But interesting plant products were there in Australia to be discovered. My later mold products were also mostly known substances that had defeated others in structure determinations or in biosyntheses.

What Are Natural Products?

The organic chemist's natural products are secondary metabolites of plants and microorganisms that mark the divergences between their biochemistries and environmental niches. In contrast, common factors such as enzymes, coenzymes, structural materials, and so forth are of interest to the true biochemist. The definition is elastic, however, and includes an increasing knowledge of the true biological functions of many such natural molecules as pigments, insect attractants and repellents, antifungal agents, UV protectants, and other survival factors. Such natural-product molecules were usually simple enough to be amenable to structure determination even by the rather primitive methods used before the 1970s. However, the structures of biochemists' larger molecules like peptides and nucleotides, as repeating units of limited structural types, are now coming into the chemically exact classes as the result of new experimental methods.

Our classical natural products, typically extractable by organic solvents, made available interesting but not too complex molecular structures as exercises for chemical investigation and training. They are often found in optically active form. Many such substances were of interest in connection with human applications as dyestuffs, pharmaceuticals, tanning agents, psychedelic agents, and so forth. Classical structure determinations (as practiced until about the 1960s) are at maximum difficulty with such initially totally unknown natural products. Nevertheless, as shown later, biosynthetic guesses could often be made about many structural types.

These difficulties and opportunities added to the personal game-playing interest, which is an important motivation for much scientific work. It is not surprising that the major methods of exact structure determination of molecules with a complexity up to about 1000 Daltons were largely evolved to cope with such compounds. Their investigation revealed much basic chemistry, and later biochemistry. Their natural occurrence also sometimes made complex molecules available in large quantities for further chemical or biochemical transformations into useful substances, short-circuiting total synthesis in areas like sex and cortical hormones, as noted later.

Classical Structure Work

The principles of structure work then technically involved two aspects: to detect and interrelate functional groups by chemical means, and to obtain structural information on the atomic nature of the main nucleus. Until the mid-1970s new substances could only be examined for their chemical transformations. When possible they were finally converted into previously known, or synthesizable, simpler compounds (e.g., by fission of unsaturation and by dehydrogenation into aromatic substances, for which synthetic methods were efficient). The interpretation of such work was facilitated by the efficient indexing of known compounds and their properties. For this and other reasons, organic chemistry is the best documented science (the order of a million substances on record).

The difficulties of this classical approach diminished as records of comparison substances continued to accumulate. Until the advent of effective physical methods (notably NMR and MS from about 1970), structure determinations could be prolonged exercises with the intellectual fascination of problem-solving on an exiguous base. One prominent chemist of the period commented to me that a triterpene structure had previously been good for two or three Ph.D. theses; now (1970s) you were unlucky if you could not solve it in a few days! More problems were needed to keep collaborators busy, but more and more complex ones could be tackled.

X-ray crystallography is increasingly efficient for determining structures of appropriate compounds, with both good and bad outcomes. The good aspect is that the results can be applied rapidly. The bad part is that less information on reactivity is now gained through chemical investigations, which were historically critical for many advances in general organic chemistry. The scientific reward system should be changed to give less credit to those who first determine a structure (no matter how) and more to those who find interesting, useful, and instructive things to do with the substance.

Personal Approaches to Research

Each investigator has personal predilections about which methods to choose from the many that are available to investigate a new structure. The basis of choice might be twofold: what had previously been noted in the literature for similar cases, or new methods suited to personal experiences and preferences. Before about 1960 the methods were all based on bond-fission processes forming simpler structures, often dependent on reaction of unsaturation, or hydrolytic fissions. Classical double-bond fissions by oxidation did not usually appeal to me because of their often unspecific and rather drastic nature, leading to further conversions of products (e.g., lanceol and curvularin, discussed later). I did not like dehydrogenations; although helping to reveal molecular skeletons, they usually obscure stereochemistry and often obscure the position and nature of functional groups.

Fashions; Competition

The structure-determination area was extremely competitive in the classical period, from about 1945 to 1970. The types of compound investigated ran in fashions, favoring alkaloids, then oxidized triterpenes, later sesquiterpenoids with new skeletons, and others. I never went with fashion. I only looked seriously at one alkaloid (echitamine[10]) and one triterpene (flindissol[11]) in my career, both of them because they were available, because they seemed unusual, and also because some students wanted to be "in the swim". I found contemplating triterpenes just as structures rather boring, apart from biogenesis.

I had one unfortunate attempt with the complex orange bitter principle, limonin, in Sydney, 1953–1954. I undertook the project because my collaborator (Bruce Chandler) had access to an infrared spectrometer in CSIRO in Sydney because the topic (bitterness in orange juice) was of interest to them. I could not use the instrument on any other topic of my choice. I agreed to exchange information with an international group and I did so freely. What I eventually received in return was a copy of a paper that had already been submitted for publication. The competition was as bitter as the substance itself, which was renamed amarin in one of the laboratories so that visitors could not understand the labels on the bottles.

Derek Barton, when the Chemical Society printed lists of papers submitted for publication, used to give uninformative general titles so that competitors in places where publication was fast could not be warned and anticipate him. Although no doubt science gained, I did not like this "battlefield" approach and always preferred a rapier to a mace. I did not have the big battalions needed for major competition. I am pleased in retrospect that I did not waste much time on structure work, which can now be done so much more easily. That raises the question of how much science gained by not waiting until methods caught up with outstanding problems.

After about 1949, I had a compulsion to work out new methods based on mechanistic understandings and techniques derived from the metal–ammonia work. However, my first structures (1937–1938) were necessarily determined through the classical approaches of the time.

Some Specific Structures

It is amazing in retrospect that, with the primitive practical separation methods available in 1937–1938 for complex mixtures of closely related C_{10}-terpenes from plants, workers like Otto Wallach and F. W. Semmler had previously managed to identify so many different new but closely related structures. One reason is that different mixtures occur in different plants, in some of which one component is dominant. Intertransformations also assisted deductions of structure in a wide biogenetic series such as this.

Terpenes

α-Phellandrene. Back in 1937, with my first major research on Earl's terpene hydrocarbon–oil mixture on my bench, my first personal question was, What is the problem? For Earl the problem was probably what to do with this waste product. For me, as a budding scientist with no prejudices in terpene chemistry and no advice, I translated this, before I could set up any program, into "What exactly does this material contain?" Distillation was then virtually useless because the equipment I had could only separate compounds with boiling points 10–15 °C apart. Perhaps I could isolate solid derivatives? This fertile type of procedure applied by earlier workers depended on purification and recognition through that efficient molecular-sorting process, crystallization. I had identified my schoolboy limonene as its solid tetrabromide and nitrosochloride. Phellandrene (1) had been characterized in the literature as a crystalline nitrosite by the action of N_2O_3 on the trisubstituted double bond.

How much phellandrene (1) does this oil actually contain? A conventional approach might have been to try to improve the yields of the nitrosite to a quantitative test. In view of product instability and its unpleasant nature (its decomposition products made me cry), I rejected that idea. Instead I went back to first base by considering the structure of phellandrene as a cisoid conjugated diene (1), which had already been shown to react under very mild conditions in Diels–Alder fashion with maleic anhydride to crystalline 2. On this basis, my obvious procedure[12] was to react it quantitatively in the water-soluble solvent acetone and to isolate the crystalline adduct by extraction with alkali as a water-soluble salt (after concomitant hydrolysis of the anhydride). Moreover, even if the adduct were not entirely chemically uniform (it was not), the unreacted hydrocarbon residue could be estimated and should contain the other non-cisoid diene components.

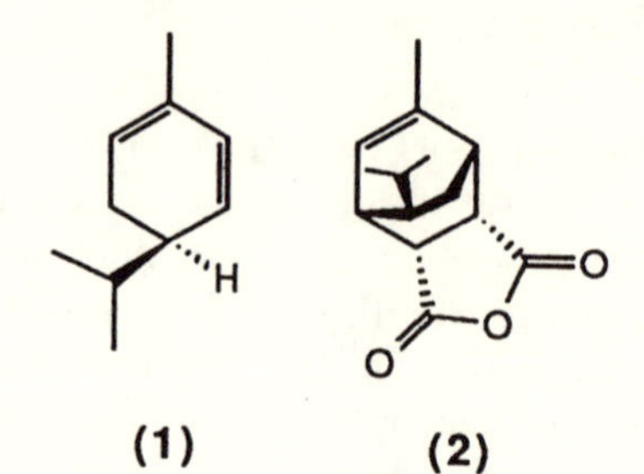

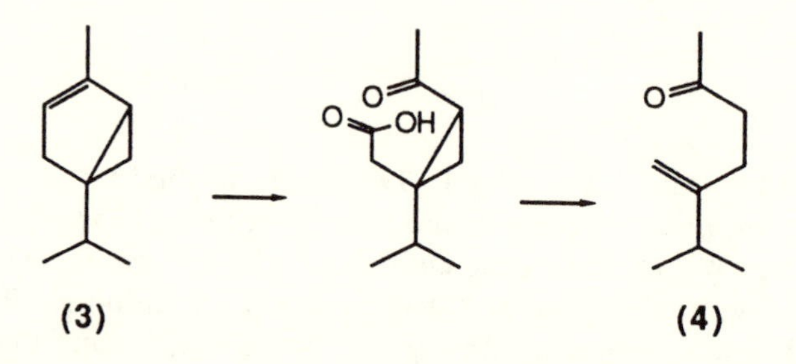

In consequence of this thinking I was able to carry out quantitative analyses that were not achievable with the initial mixture. I showed from the crystalline adduct that the phellandrene was fully optically active (which could not be deduced from the mixed crude oil) and that the residue of the oil was mainly higher-boiling aromatic *p*-cymene, together with α-thujene (3). With a lower boiling point than phellandrene, α-thujene was now separable by distillation in a reasonable state of purity.

α-Thujene. My isolation was the first identification of natural α-thujene. Its cyclopropane ring conjugated with a double bond gave me ready access to a unique structure provided by Nature, not then available by synthesis, for carrying out new chemical transformations initiated on the double bond. This situation recalls the availability of α-pinene with its double bond conjugated with a cyclobutane ring, the chemistry of which led to the first understanding of carbenium ions and cyclobutanes. My substance provided the material for a number of novel synthetic transformations. Thus, in Earl's absence, four of my first five papers were in my own name alone.

α-Thujene (**3**) illustrates a primitive procedure for structure determination. Oxidation of the double bond with permanganate splits this compound to yield a keto acid that I did not recognize initially, although it had been described in the literature. Because my sample was not fully optically resolved, its properties were not the same. Distillation, in an attempt to purify it, led to decarboxylation to a known optically inactive ketone **4**, which was recognizable from its mode of generation, and a description in the literature of this now optically inactive product.[13]

A Lignan: Gmelinol

I simultaneously worked with Lions on the structure of the natural lignan gmelinol (**5**),[14] which was extracted with boiling water from masses of wood shavings of *Gmelina leichhardti*. Lions recognized the opportunity when he saw crystalline material in the timber. Many early natural-products chemists depended on similar direct observation. Lindsay Briggs (New Zealand) used to tramp through the bush smelling and tasting. His claim to be able to recognize worthwhile compounds immediately was witnessed by his results. My structure work was not facilitated by Lions's irrational conviction that the molecule is C_{19}- and not C_{20}-, as my analyses clearly indicated. (It is a lignan, as I shall discuss later.) This compound, and its acid-catalyzed conversion (oxonium ether) products, such as the stereoisomer isogmelinol (**5a**) supplied me with amusement for years. During this period our structure-determination methods advanced from stage to stage, culminating in the ability to correlate NMR spectra with structure and stereochemistry in the whole tetrahydrofuranoid lignan series.[15]

My first structural attack on gmelinol was based on brutal classical methods: Destructive distillation to 1-methyl-3,4-dimethoxybenzene and the corresponding aldehyde gave the clue to the benzylic ether (veratryl) structures present. Oxidation to more than one mole of 3,4-dimethoxybenzoic acid demonstrated the existence of two of these in the molecule, but the C_4O_3 piece between them was mysterious. Acid-catalyzed isomerizations at the ether centers gave new, unnatural stereoisomers.

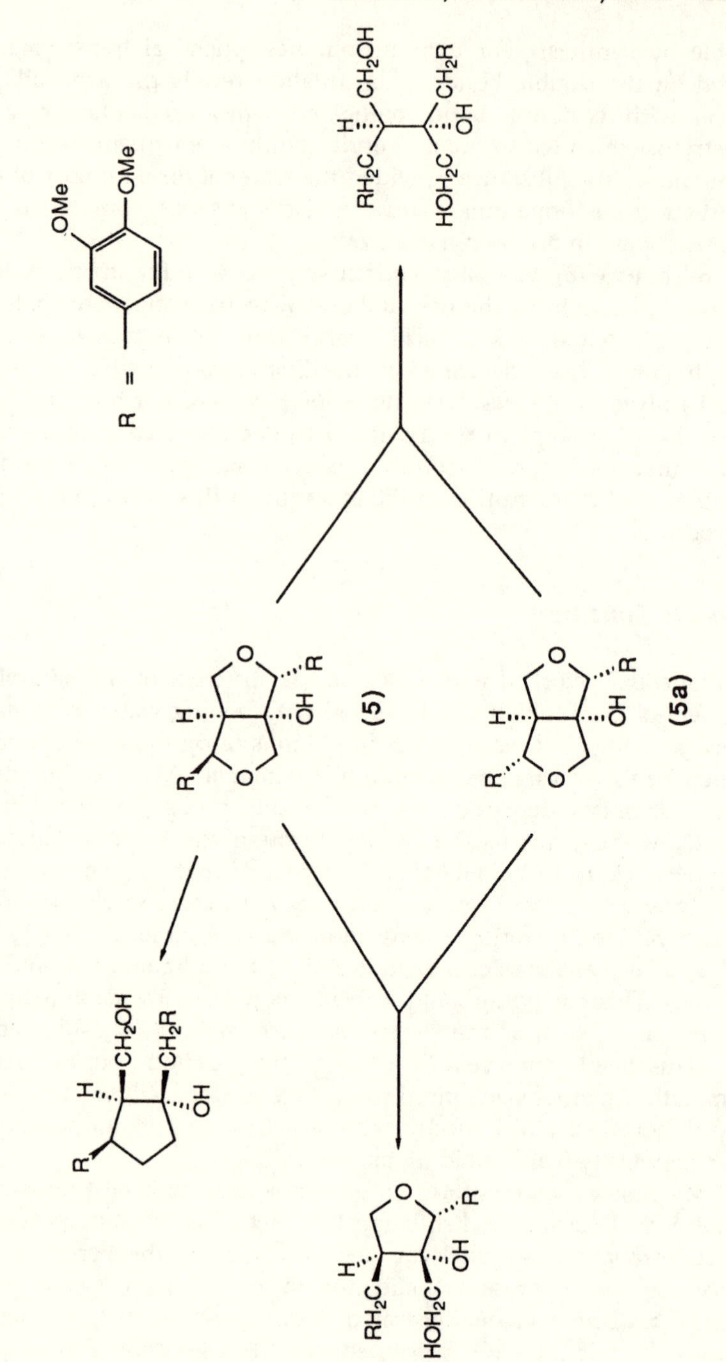
R =
OMe
OMe
(5)
(5a)

Personal Predilections for Carbanions in Structure Work

As I showed,[16] the key to determination of both structure and stereochemistry for **5** and **5a** (summarized in the attached formulae) was based on the selective metal–ammonia fissions of the ether groups at the benzylic situations (through carbanion intermediates). In each isomeric case one asymmetric center was experimentally abolished at a time. The initial structures present were then interpretable from the numbers of the isomeric definable structures resulting in each case. The conclusions were supported by the equilibration steric positions of products through oxygen protonation. The overall principles of such deduction were the same as those used by Kekulé 100 years earlier in arriving at *o*-, *m*-, and *p*- substitution patterns of benzenes from the number of isomers deducible by theory.

Similar fissions and equilibrations permitted our first determination of the relative and absolute stereochemistry of the important tannins, catechin and epicatechin.[17]

"Fingerprints" of thinking are revealed by broad approaches, and my types of structure determination show how results in any area fertilize applications in another, given the right mental approaches. Many methods I developed for structure work depended on metal–ammonia fission reductions or base-catalyzed reactions using KNH_2–NH_3 (to form carbanions or as NH_2^- donors). The methods were originally devised for synthetic needs, but lead to fundamental mechanistic understandings.

The reduction reactions used in my work from 1944 to 1950 (Research Set 2) demonstrated, rather unexpectedly, that metal–ammonia solutions could produce C–O fissions in closely defined structural situations (benzylic or allylic). An intermediate in each case was almost certainly a mesomeric carbanion, which could explain the substitution requirements and products. These results led to an understanding of the formations and reactions of general carbanion structures, including those resulting from other base-catalyzed procedures. I used these procedures in other appropriate structure cases, such as with alphitonin **8**, R = H), neocembrene (**9**), and geijerene (**17**), as discussed later.

Simplifying Structures

I also applied my early experimental and theoretical conclusions on reduction-structure specificities to structure work in several ways that were new in principle. My approach was to simplify (in a rational manner by reduction with C–O cleavage) a natural structure that contained too many oxygen substituents to be dealt with by classical methods. An alternative would be to form one that could then yield a product recognizable from the literature (e.g., methylgeraniolene from mycelianamide).

New, even if unknown, reduction products are susceptible to simpler determinations of structure than their parents are, including the obliteration of asymmetric centers. Knowing the structural course of my reductions, I could deduce a few alternative structures for the initial natural oxygenated precursors and test the postulated structures by other methods. A major structural feature of the chemical reductions useful in this connection is that, in contrast to catalytic hydrogenation methods, unconjugated C=C are unreduced and usually remain in place. Even if they migrate during reduction through a (predictable) mesomeric carbanion, original situations can be assigned on a mechanistic basis.

To exemplify one aspect, I refer first to the sesquiterpene alcohol lanceol (**6**, R = OH), which had been assigned an extraordinary structure by Simonsen following his misinterpretation of a standard oxidation procedure. His classically assigned structure was unbelievable to me as a terpenoid. His reaction had resulted in an alkaline solution, which caused secondary changes that he did not recognize. Using a probable initial biogenetic sesquiterpene structure, I could easily formulate lanceol on paper in conformity with the known scanty chemical evidence as (**6**, R = OH), but how could I test that idea? There was no doubt from Simonsen's work that the structure contained $MeC{=}CCH_2OH$, so I reduced this molecular structure by $Na{-}NH_3$ predictably into $=CMe_2$ and obtained my foreseen and readily characterized hydrocarbon (**6**, R = H).[18] This structure took about 3 hours of work in the laboratory, with intuition foregoing further perspiration.

A similar but more complex case was the plant glycoside aucubin, which I do not discuss here in detail. Its aglycone is unstable to the acidic hydrolysis necessary to remove the sugar for structure examination. Chemical evidence was available that it has at least one allyl alcohol structure present. Metal–ammonia reduction of it, and of its acetyl ester, showed that there are two such structures (in different molecular environments as indicated by the differing ease of reductive oxygen removal in stages, which could be structurally correlated from our previous general work with the extent of double-bond substitution). The chemical formulation of the reduced aglycones led us back to a correct formulation of aucubin,[19] independently reached by others on different grounds.

Support of Biogenetic Structures

Clear definition of biogenetic relations can assist structure work, as I shall discuss later. Todd asked me in 1949 to look at the natural anthelmintic protokosin.[20] From what little was known, it was clearly chemically related to the male-fern anthelmintics such as flavaspidic acid containing two rings, a series then still incompletely formulated, and on which I was speculating biogenetically from published evidence. At first I sought from the literature what I judged to be simpler monocyclic natural models. Many of these had not been formulated at all, or wrongly formulated, but I recognized a fundamental structural series related to alkylated acylphloroglucinols, which I classified as enolic β-triketones.[21] Among these compounds were the Australian essential components xanthostemone and tasmanone (7), of which I similarly derived the structures.[22]

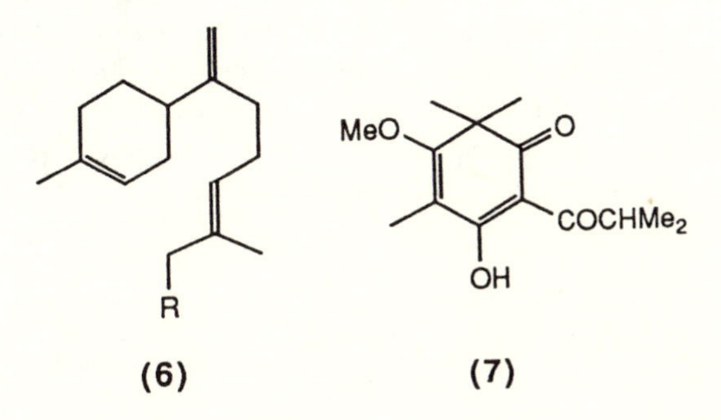

The major difficulty then with structure determinations in this series by chemical methods was that most of the substances have either oxygenated functional groups such as carbonyl or double bonds on every second carbon of their six-membered nuclei. Standard oxidation or alkaline hydrolysis methods therefore uninformatively pulverize the molecule. I asked myself if reductions could be used instead.

The experimental answer was yes. I found the carbonyl of an acyl side chain in such structures to be selectively reducible, as I predicted, to an alcohol (by Na–Hg). The product then was split by base-catalyzed retroaldol fission, in a predictable manner, into two readily recognizable fragments: the side chain characterizable directly as a volatile aliphatic aldehyde, and a simplified, recognizable, aromatic nucleus. This novel direct conversion process could be carried out experimentally in one stage through reduction followed by steam distillation. The method ultimately permitted assignments of structure to many compounds,[21,22] including (by extension of ideas) to the important, more complex *Filix mas* anthelmintics of then incompletely known structure. The alternatively C- and O-methylated phloroglucinol structures found in this series were shortly to be important intellectual ingredients in my then-developing (1951) biogenetic speculation on Me cation introductions (Research Set 3).

Alphitonin: Amide-Anion Attack. Another structure case with too much oxygen for standard oxidation and hydrolytic methods was alphitonin (**8**, R = H). While I was seeking a highly specific (basic) fission process to split it into informative pieces, it occurred to me that if the OH were "covered up" as OMe (**8**, R = Me) to prevent salt formation, my biogenetically suspected (flavonoid-related) formula would contain a nonenolizable ketone. In principle, this ketone should undergo C–C fission by attack of the amide anion adjacent to the carbonyl and break into simpler fragments (the Haller–Bauer-type fission that I had used in my early Oxford syntheses). Elimination of an aromatic carbanion (characterizable finally as 1,3,5-trimethoxybenzene, facilitated predictably by the OMe) and a simpler recognizable fragment should result as shown. I experimentally greatly improved yields in the fission process through KNH_2–NH_3 (liq), which I was developing for other purposes, thus confirming the structure.[23]

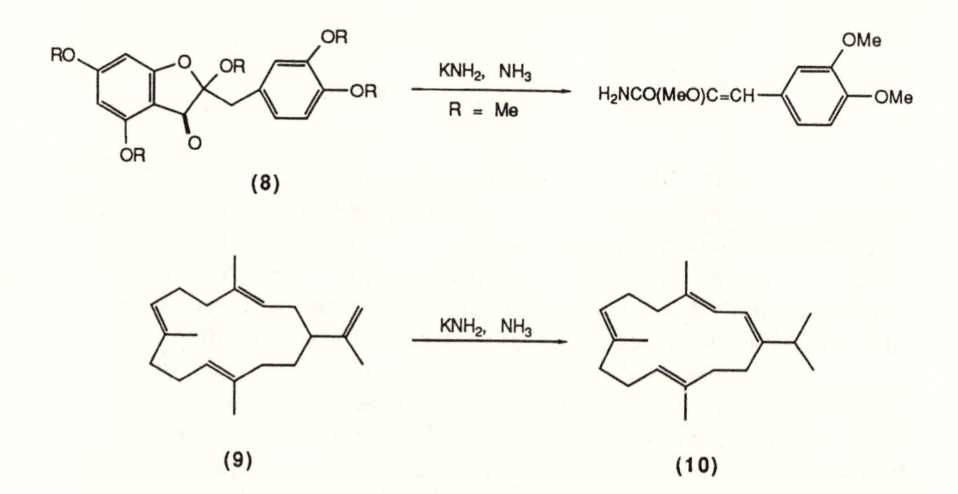

Neocembrene: Base-Catalyzed Double-Bond Migration. An altogether different type of natural structure was that of the termite food-trail pheromone, the diterpene hydrocarbon neocembrene (**9**). We examined this substance in Canberra in the early 1970s when we had access to developing physical methods, but their use and interpretation was greatly facilitated by a specific base-catalyzed double-bond migration process with unusual structure specificity, the course of which I could predict from my early synthetic metal–ammonia work (Research Set 2).

My initial work[24] had shown that KNH_2–NH_3 could produce double-bond migration only in a defined substitution type, for example, moving the double bond of $CHC{=}CH_2$ into a chain (including for the first time allyl into enol ethers), but it did not affect more highly alkylated double bonds. It could even more readily bring C=CCHC=C catalytically into conjugation. Combined with biogenetic speculation (an isoprenoid diterpene overall structure from the C_{20} formula and degree of reduction?) from NMR spectra and a new quantitative microozonolysis–GLC procedure for estimating recognizable products of double-bond fissions, we were able to first guess at **9** and then confirm it by the transformation shown into **10**,[25] which was convertible by double-bond fission into a known product from terpinene. Only a several-milligram quantity of a liquid compound was initially available (from 20 kg of termites) and most of it remained afterward.

Personal Satisfaction

One of the fascinations of this sort of natural-product work is that you never know what sort of novel structure will turn up, and what its general chemical and biosynthetic implications might be. I am sure that I obtained more personal satisfaction from such novel approaches than if I had only been interested in the structures as such and had had available the present powerful but boring spectral methods. A lot of reaction chemistry was learned, which would have been missed with the present methods.

Other Interesting Structures

Among 100 or so structure determinations with which I was associated, I can point out only a few general features of broad interest, reserving important biogenetic ones for Research Set 3. Much of my interest in determining new structures was in their biogenesis or biological activity, but some attracted me because of their obviously unusual character.

Neocembrene (**9**) was the first observed occurrence of the primitive biogenetic precursor for the whole macrocyclic diterpene cembrenoid series (one of the two major bases of the diterpene area, as I discuss elsewhere for gibberellins). Aromadendrene (**11**, R = CH_2)[26] was the first sesquiterpene structure with a confirmed cyclopropane

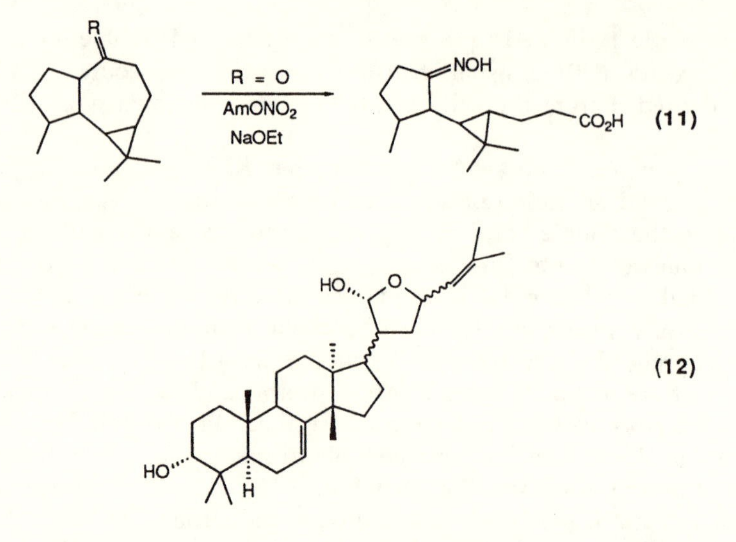

ring, determined by a novel process as shown, and predecessor of a major series. Flindissol (**12**) was the first example of the type of biogenetic missing link lying between a complete triterpene and the class of limonoids (bitter principles) and other triterpenes with a furan ring replacing the initial biogenetic side chain. Its structure also suggested to us a biogenetic mechanism for the natural side-chain degradation to form an aromatic furan.[27]

Some Unusual Reactions Interpreted Structurally

Antimycin-A

Some interesting types of structure occur that are not initially known in synthetic substances, but which then provide targets for synthesis and for novel reactions. The fungal antibiotic antimycin-A (**13**, with various aliphatic side chains), first attracted us. What was known of its structure from chemical evidence suggested it as one possible vehicle for our biosynthetic studies on branched aliphatic structures. But as a basis for this we were led to suggest the structure of its novel dilactone nucleus, the steric compression of which rationalized its extraordinary alkaline fission reaction that was shown to generate a neutral lactone. This unusual type of process was not then known in any synthetic structure but could be conceived of as a consequence of steric compression in a medium-ring dilactone precursor.[28] Our proposed structure was later confirmed by van Tamelen et al.[29]

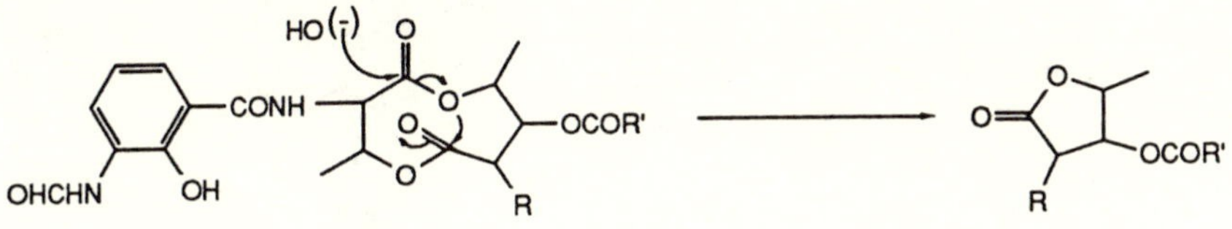

(13)

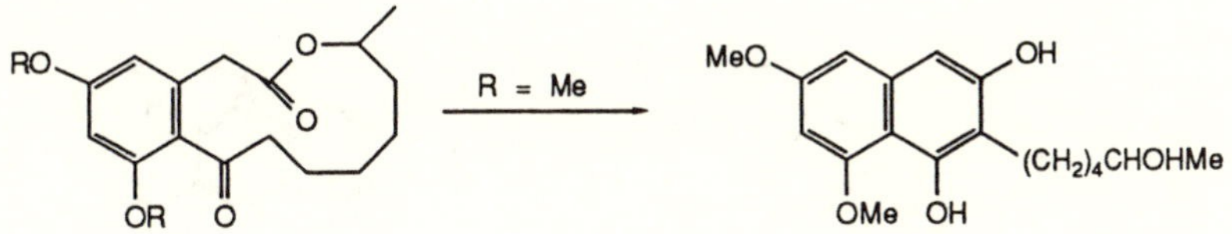

(14)

Curvularin

We also formulated the mold metabolite curvularin[30] (**14**, R = H) on the basis of the extraordinarily facile alkaline Claisen cyclization of a lactone in water at room temperature into a fully aromatic naphthol derivative. These experimental conditions differ dramatically and significantly from normal laboratory requirements for very strong bases in similar Claisen ester closures with usual structures. We were forced to postulate a novel facilitating, compressed medium-ring structure.

The ease of this laboratory process suggested to us that analogous medium-ring intermediates might in general lie on biosynthetic pathways into many polycyclic aromatic polyketides (Research Set 3).

Mechanistic Prediction of Structure: Leucoanthocyanins

Another approach to structure was prediction, on the basis of reaction mechanisms, of a biosynthetic precursor type required to form a known type of natural product from suspected precursors. This approach led to my suggestion in 1953 of the structural type (**15**, R = *p*-OH phenyl and related systems) for the leucoanthocyanidins (proanthocyanidins),[31] transformed with acid in air as shown into pigment anthocyanidins.

It was our first speculative biomimetic synthesis of this reaction-based class. The reactivity basis was the expectation of removal of a 4-OH (from reduction of the 4-carbonyl necessarily present in my view in its biogenetic precursors) via a benzylic carbenium ion. The class turns out to include our suspected biogenetic intermediates in the conversion

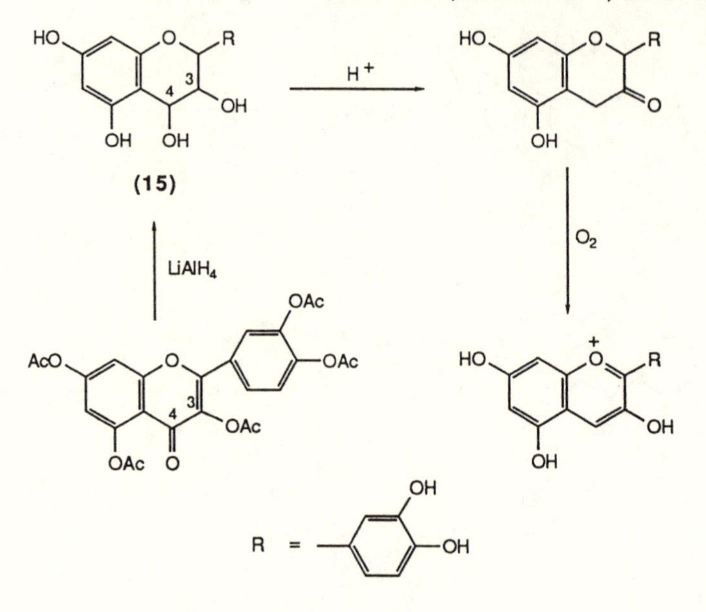

of flavonoids into the anthocyanin pigments of flowers (for further details, *see* Research Set 3).

Early Physical Methods

Nowadays, the use of advanced NMR techniques, mass spectrometry, and X-ray crystallography often renders the older chemical methods superfluous, provided that only structures are needed. A structure can be derived from small amounts of sample, sometimes without even seeing a compound. This development leads to a great loss of chemistry learned from unique products. I have been trying to persuade my Southeast Asian colleagues (in the S.E. Asian Natural Products Network, which I helped to found) among others to seek to promote the pursuit of interesting new chemistry on available known natural compounds[32] to make up for this modern lack.

About 1943 I used ultraviolet spectra for structural purposes (for example, showing to Robinson's astonishment that a synthetic product is an α,β-unsaturated ketone, although it did not give a standard 2,4-dinitrophenylhydrazone.[33] I then used a laborious Hilger photographic technique. The general idea came from my acquaintance with the work of A. K. Macbeth (Adelaide), who used ultraviolet spectra with terpenes in the 1930s. He told me he had learned it from A. W. Stewart who was head of chemistry in Belfast.

In the early 1940s, H. W. Thompson (PCL, Oxford) developed double-beam instrumentation for IR spectra. I gained access to this "black-box" technique through knowing several of his collaborators and used it chiefly for analytical purposes on the mixed reduction products from my first metal–ammonia reductions (about 1943–1944). In 1954 I visited R. A. Ogg in Stanford, who told me exactly what NMR would do for the organic chemist, and he was right. However, no "black box" was then available. I have never been a developer, or even an expert user, of complicated instrumental techniques, a characteristic I greatly regret. I merely filed those ideas for reference, as I shall discuss later.

Zierone

I did use UV and IR spectroscopy when possible, although I had no instruments of any kind available in Sydney (1952–1955). One example of the correct use of UV then (carried out in Oxford by my friend the analyst Strauss) was to help define the substitution pattern (2,4,8) in the azulene, zierazulene,[34] which was derived by dehydrogenation of the sesquiterpene terpene zierone (16) (of then unknown structure), a natural ketone kindly given to me by Maurice Sutherland. Its reactivity indicated a ketone, and probably an α,β-unsaturated one. I sent it to Strauss in Oxford, asking whether there is an absorption maximum somewhere in the 230–280-nm region. His correct, but misleading,

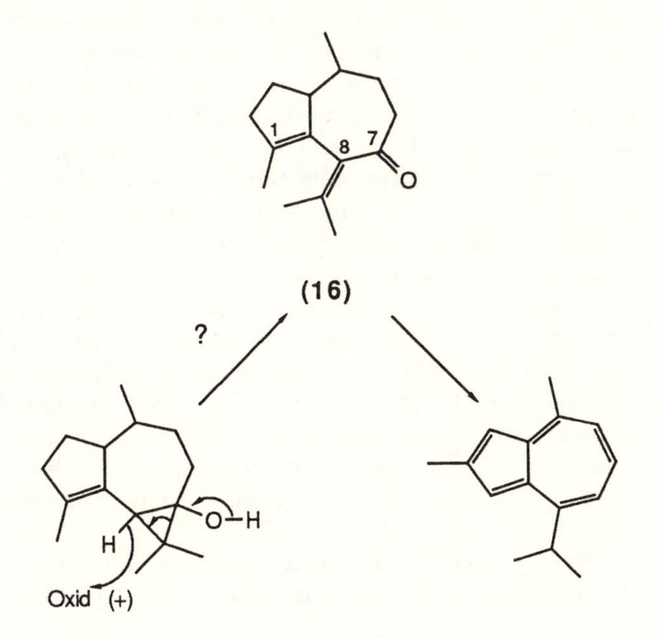

answer was no. Compelling chemical evidence for an α,β-unsaturated carbonyl emerged later. When I secured access to an instrument to examine the spectrum myself, I found high absorption (ϵ about 5000) in that region, although no maximum. I had asked the wrong question.

The problem with interpretation of its spectrum was due to lack of coplanarity of the carbonyl and double bond because of steric interference of an adjacent 8-$Me_2C=$ and a 1-Me. This interference also explained the extraordinary ease of migration of the 1-Me into the 2-position during dehydrogenation, forming a final purple rather than blue zierazulene. Its color probably indicated 2,4,8-substitution rather than the correct 1,4,8-substitution. This steric interference was a major complication, until its structural basis was recognized.

Such transformations and spectra provide another example of the value of natural products in demonstrating reactions and properties of synthetically inaccessible structures. A new biosynthetic pathway as shown for **16** starts with my biogenetic aromadendrene-type **11** skeleton, explaining the steric strain.

Mass Spectrometry and NMR Spectroscopy, an Experiment in Physical Methods: Geijerene

By early 1956 (newly in Manchester) it was obvious to me that the limitations of existing spectral methods needed supplementation by methods to define large parts of the molecular skeleton. Ultraviolet and visible spectra were informative in limited ways about unsaturation, infrared spectra about certain functional groups and their environments. John Watkinson and I tried in 1956, in Manchester, to develop with high-resolution IR what can now be readily done with NMR, but we failed because of the complexity and lack of computers. Mass spectrometry was then the only physical method I could think of that could potentially give information on large skeletal fragments. X-ray crystallography was then rarely usable, very laborious, a physicist's technique, and by definition required crystals.

I spent a week in bed with influenza in 1956, with the Petroleum Handbook on MS, a technique then used solely for identification and analysis of mixtures of simple hydrocarbons. It was a bad choice of information (the only one I had), because of the lack of functional groups in the substances and because gas-phase reactions and stabilities were strange to me as a solution chemist. Nevertheless, I decided to try an experiment using NMR and MS to determine a natural-product structure.

This project related to an arrangement I had made with Maurice Sutherland (University of Queensland) when I left Australia. He had given me the C_{12}-geijerene (17), of then unknown structure, initially as a collaborative project. When I left Australia he was understandably unwilling to give me full control of his compound in England. We arranged that he would examine the structure by classical methods, the only ones he had, while I in Manchester confined myself to new physical methods. In fact, the only important reported chemical reaction I repeated was hydrogenation, to confirm by MS the number of double bonds.

Also, following ideas generated in discussion with Ogg in 1954, I was able to arrange with Norman Sheppard about 1957 to have an NMR spectrum of geijerene run on the first instrument in Britain. This produced a minute photographic trace, read with a lens, but structures like $C{=}CH_2$, $CH{=}CH_2$, and quaternary Me stood out even to an amateur like me.

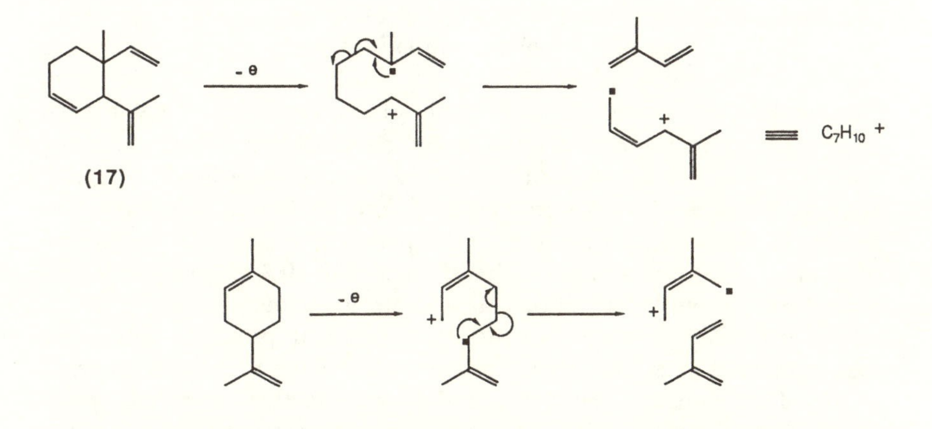

Mass spectra were run on a spectrograph belonging to J. H. Baxendale in Manchester, who greatly restricted my usage to low molecular weights (which he thought he might clean out of the machine). I could not obtain another instrument because there was one already. We were able to interpret mass spectra in terms of 17 as shown,[35] although only after reducing the energy of the electrons to about 10 eV. The principal fissions, supported by our work on limonene as shown, were interpreted as a reverse Diels–Alder reaction and loss of quaternary allylic Me.

As a confirmatory exercise, I also isomerized structure-specifically the double-bond system of geijerene with KNH_2–NH_3 in a manner similar to that I later described[25] for neocembrene. I read a paper on the work at the 1960 Natural Products IUPAC in Canberra. Apart from K. Biemann, who discussed a synthetic peptide, nobody had yet realized the potential of the MS approach for structure determinations, although it had been extensively used for analysis. Djerassi chaired the session and later, in 1962,[36] began to investigate mass-spectrometry in a major fashion. Geijerene (17) was the first natural product for which the structure ultimately rested on physical methods alone, despite some earlier chemistry.

I did not pursue mass spectrometry for several reasons. One was a prolonged dispute about publication. Maurice Sutherland had reached the same conclusion independently by chemical means and directed the Chemical and Industry Society London proofs of his paper to me in January 1960, as a signal to publish simultaneously. Unfortunately, the proofs were corrected by Herchel Smith during my absence in Nigeria, and he did not mention this to me because he did not realize their significance. When I submitted a paper I was accused of trying to steal undue credit, despite the attested evidence. One referee, Paul de Mayo, whom I recognized by style (it had to be he or Derek Barton) and challenged was particularly obdurate. I had to appeal to the Publications Committee of the Chemical Society. His final comment was "Well, you got it published eventually, what are you complaining about?" The key word was eventually. The editor, R. S. Cahn, who did not like me (I objected to his uninformed rewriting of my papers, including renaming my 19-nortestosterone as 10-nor, after the proof stage of publication, without informing me), also insisted on many changes, like the title, "An Independent Confirmation...". It was either one or the other. My title was "An Experiment in the Use of Physical Methods". Cahn once told me that if he cut out 10% of each paper submitted to the journal he earned his salary. Rejecting one was, I guess, even better. My comment in good Australian is unprintable. My paper was published in much delayed and attenuated form,[35] which removed much of its impact. The date on the paper is that of its editorially imposed modification, not its original submission. I was astounded that slanting accounts of origins of ideas to steal credit seemed assumed to be the norm. I had to prove my innocence several times over, whereas faking actual experimental results is the greatest scientific sin and usually is assumed not to occur.

A difficulty in continuing the work on a general basis was that Baxendale would not let me put anything into his MS machine that he thought would not easily come out (understandably), so the possibilities of pursuing natural-product mass spectrometry were for me limited to very small molecules, which were not the important problems. I tried without success to get R. I. Reed and John H. Beynon interested in this structural aspect as collaborative ventures.

Later I could have been able to cope administratively with such political situations, but not then. Other workers, notably Djerassi[36], had made rapid progress after instrumental capabilities and attitudes were clearly ripe for this development. I was never one for personal confi-

Carl Djerassi in the mid-1950s. Djerassi has written two autobiographies, Steroids Made It Possible *and* The Pill, Pygmy Chimps, and Degas's Horse.

dence with complex instruments. Also, I was busy initiating other academic areas of work. With industry (Associated Electrical Industries), we later designed an inlet system for the MS9 that did not pyrolyze samples. I later used physical methods freely, but had no personal interest in their development as instrumental techniques.

Behind Every Successful Man ...

I should pause at this stage to look at two women who made possible the beginning and the continuation of my career. As an only child I was rather a loner, with few but firm friends. This group did not include girls, who were equivalent to creatures from Mars until I went to the university at 17. My adolescent energies even then were almost necessarily directed toward scientific interests because I had no money. I also experienced only a few years without major illness in my family between 1937, when my father died, and 1942, when my mother became ill.

Despite my general social alienation, several women proved extremely important in my career.

Lily Bailey and Jessie Williams

I owe my career first to my mother, Lily Bailey. She was 38 when I was born. She had an interesting background. Her paternal grandfather, John Bailey, was a convict transported to Tasmania about 1822 for "robbery from the person". Her grandmother, Dorothy Woodhead, was similarly transported for "stealing linen". Obviously, they were enterprising in dealing with the social problems of their time. John made good, becoming a large landowner in central Tasmania. On the other side of the fence, her maternal grandfather, Denis Thomas Maher, was a quarrelsome Irishman who apparently ran into trouble in the British Army in India, finishing in the garrison of the convict settlement at Port Arthur in charge of the commissariat. He traditionally drank himself to death on Irish whiskey. The Baileys were friends and neighbors of the influential Page family, who ran the first coach service in Tasmania. My grandfather, also John Bailey, was born in their "Bath Inn" coach stop in 1846. My mother was born in 1877 in the cold and lonely heart of Tasmania, at Lemon Hill near Oatlands, and had to ride 7 miles to school on a horse. She finally rebelled at 27 and emigrated to New Zealand, where she met my father.

My mother was one of the kindest and strongest people I have ever known. She supported my ill father for 7 years (no pensions) and encouraged me in the critical years 1933–1935 to fulfill my ambitions at the university, although my father, understandably, wanted me to earn a living. She gave me a room and what little other support she could. She claimed to be stupid; in fact she was highly intelligent, but not intellectual. I still have her school prizes to prove her ability.

After she followed me to Britain in 1939, I was her only connection. In 1942 she showed signs of Parkinson's disease. If my career had to suffer so that I could look after her, so be it; I could not abandon her.

My mother, Lily Bailey Birch, and I on Thirroul Beach about 1917.

With my mother (left) and Aunt Madge (one of her seven sisters) at grandfather's farm , "Princethorpe", in Tasmania about 1919.

I knew that if she were put into a nursing home she would die. This ultimately happened in 1951, when I could cope no longer. Her sickness explains my long stay in Oxford, although I did not regard the situation as scientifically favorable. I could not in conscience ask any girl to take on both me and my sick mother, although the situation was assisted by her enormous vitality. Also, in a way, the War helped my personal battle. If I had to queue up for limited rations, so did everyone else. If I were physically restricted, so were they. My mother was happy to see me achieving my goal. Thus the results of my work became the sole redeeming feature of a sad situation. If I failed it was a waste of her endurance and mine, so the emotional pressure to succeed was tremendous. I was "putting money in the bank" for some problematical future, when there would be "blue birds over the white cliffs of Dover". Some women complain, to explain failures, that such situations are thrust upon them. But the situation does happen to some successful men, and I was not married until I was 32, so was on my own as a carer. The outcome is a matter of attitude of mind: Indomitable is as indomitable does, no matter what the pain, as my mother observed.

In 1947 Jessie Williams appeared on the scene, first as a nurse for my mother. She was 25, handsome rather than pretty, so she still retains her Welsh good looks. She had started as a librarian in her native Cardiff. Then she became a nurse, training in "bomb alley" at Croydon (1941–1942) and working in military hospitals. When she tired of the "monastic" hospital life, she became a district nurse on a bicycle, first in

Jessie Williams in the early 1940s, just before I met her.

"Looking poetic" in 1947. The look was due to starvation, not poetry, but it attracted Jessie.

the notorious Soho of London, and then in Oxford. She took my mother almost literally out of my hands, and eventually married me. Her joie de vivre, her artistic creativity, and her intelligence attracted me. What attracted her I cannot say, except that I had long black hair, was half-starved, and she said she thought I looked "like a poet". The importance of her decision to marry me was immense personally; for my career, it freed me to leave Oxford. We were married in late October 1948, and in January 1949 we were in Cambridge.

Jessie and I at our wedding in 1948 in the Saxon Church (1040 AD) St. Peter in the East, Oxford.

Jessie's subsequent career, in arts organization, with her own art works, as an employee of the Australian National Gallery, and in many other creative public efforts for theater and family planning, demonstrates that she obviously could have made her own independent creative career. I always asserted that she shared my scientific achievements, but perhaps she has not been entirely convinced. The artistic world that surrounded her supplemented my scientific one with a different kind of rationality. To her I owe the continuation of my career, notably in Cambridge, but also in Sydney and Canberra. This support included her decisions, with sick children, to uproot herself to Australia, back to Britain, and again to Australia. It was a marriage made not in Heaven, but on Earth. She is now a very happy grandmother and golfer.

My mother at my wedding; all but indomitable.

Secretaries and Their Ilk

While on this topic of support, I must acknowledge the great contribution to my career of some other ladies: notably secretaries like Bev Cooper and Betty Moore, and my nonchemical research assistant Maureen Kaye (now alas dead). Their contributions probably doubled the efficacy of my science. They took from me the burdens of memory and of organization outside technical science. They organized me personally, they dealt efficiently and kindly with the outside world on my behalf (Maureen, for example, did most of the work for the first major "hands-on" science exhibition in Australia, "Australia '75"), and indeed they often knew more about writing scientific papers than I did. I expected them to use initiative within broad guidelines, and I always supported their judgments. Bev, I am pleased to say, is now in a high administrative university post, which she always merited. She married very young, had a family, and stumbled on the promotion path, but her intelligence, charm, and enterprise got her there in the long run. Their names do not appear on scientific papers (Maureen's did on a joint book), but perhaps they should have.

Acknowledging Other Contributions

I make no dispute that many people contributed profoundly to my science: not only these ladies, but technicians and laboratory managers (notably John Harper in Canberra), and academic assistants with techniques like growing molds (Margot Anderson), MS (John Macleod), and NMR (Dick Bramley). The traditional system is inadequate in the modern world because of its lack of appropriate acknowledgment in print of the fact that teams, not only original scientists, must contribute to successful research. I did my best with salaries and promotions to recognize my debt. There is no possibility that I, personally, could have performed the techniques that such people contributed.

"Administration" as carried out at the Research School of Chemistry, about 1972. From left, Jill Peck, David Craig's secretary and now a sculptor and art lecturer (we chose well); Bev Cooper, my secretary; and Rod Rickards.

Research Set 2

Total Synthesis: Novel Methods and Reaction Mechanisms; the Birch Reduction Becomes the Birth Reduction

One of the most spectacular achievements in science is the ability to synthesize complex molecular structures by rational sequences of reactions. Initially this capability followed knowledge gained from the results of random experiments. Researchers threw myriads of molecules (in any visible quantity of material) together to find out what would happen. (The W. H. Perkin, Sr., synthesis of the dyestuff mauve is a classic example.) Novel synthetic reactions leading to predictable structures in the early periods usually followed the extrapolation of such unanticipated observations. During the period in which I worked, an increasing understanding of the mechanisms of random collisions became translatable into practical synthetic methods. As examined in Research Set 4, this process is now being further extended into rational assembly methods that are reminiscent of techniques used with enzymes.

In 1941 the industrial synthesis of natural substances such as vitamin A and β-carotene seemed a challenging dream. They had not been made even in the laboratory, despite considerable efforts. The structures of steroid hormones were becoming known, but even partial syntheses were not available, and total syntheses provided a still greater challenge. However, confidence was gathering, fueled by the total syntheses of some vitamins like B_1. It was the beginning of the "classical" period in synthetic capability for defined complex structures up to about 1000 D (or higher with repetitive units).

Until about 1975 synthesis was often needed to confirm a classically elicited molecular structure. With physical methods the situation changed.

The art of synthesis, developed in parallel with growing understanding of mechanisms, became increasingly sophisticated. It is now probably the highest art of organic chemistry. Synthesis can be computerized to assist the analyses of structures for synthetic units and the availability of reactions to link them. However, synthesis is likely to remain largely a personal art for the future, involving dreams and unquantifiable estimates of what reactions and sequences will work and how well. In the 1930s synthesis was capable of dealing with simple molecules like vitamin B_1, which has two aromatic rings and no steric problems, but even vitamin A (with steric problems) was only synthesized after World War II. Spectacular successes from about 1950 (e.g., A. R. Todd and F. Bergel's synthesis of vitamin B_1) increased chemists' confidence to tackle successfully more and more challenging molecules, up to ones like vitamin B_{12} in the most complex set of classical synthetic procedures ever carried out.

Needs associated with increasingly complex molecules, particularly involving steric problems, led to sequences that required an increasing number of closely controlled stages. With a "reverse compound interest" yield at each stage, the emphasis had to be placed increasingly on experimental efficiency, particularly if product material were needed for use and not just for information or for the demonstration of personal genius. Because I required usable quantities of steroid hormones, my own synthetic approaches had to involve such strategic syntheses and I neglected alternative impractical ones.[37]

Robert B. Woodward, by using microscales, physical methods, and dedicated workers, demonstrated the power of a combination of ideas and experimental efficiency at all stages. For example, Bob told me in 1950 that his group had used my unsaturated carbonyl deconjugation ideas in attempting about 130 times, on a microscale, a key step for making ring A in his steroid synthesis, before any of the desired product was obtained. Robinson belonged emotionally to an earlier era. If he did not manage to make one simple sequence work, he tried another entirely new ingenious and simple one. I partly inherited that attitude, but I did not have either the practical means or the personal confidence to invest large efforts in long sequences that might well fail.

Early in my career I did not have to consider the needs of ab initio extended steroid skeletal synthesis. For reasons I shall discuss, I needed to devise individual unit stages to complete desired steroid hormone molecules from Robinson's starting material (**18**, **19**), which he synthesized simply as shown.

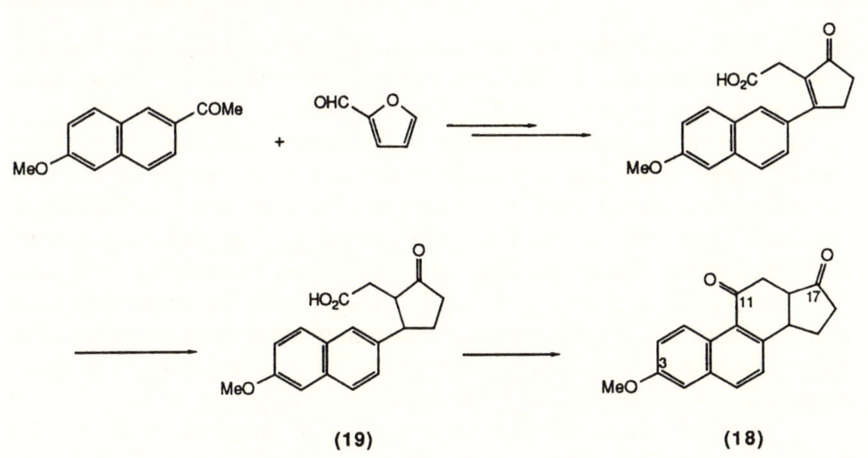

Later when I did attempt purely personal novel syntheses, a carryover from this earlier psychological approach was an inner personal demand that at least one stage had to be a new procedure that I had devised. Unless I urgently needed a compound (e.g., to confirm a structure or to make a biosynthetic intermediate), I did not follow synthetic sequences that used only known reactions, no matter how ingeniously engineered.

German Fighter Pilots and Cortical Hormones

In 1941, in wartime Oxford, I was finishing my D.Phil. work and was cut off from Australia. Although I had been very strongly anti-Fascist and anti-Nazi since about 1932, it was apparent to me and to Robinson that there were better ways than the army through which I could contribute to the war and its aftermath. He said to me in 1941, "The Americans want us to put all of our scientists into the forces or into factories, but if ever we win this War we will need some people to start up again. I want you to be one of them. I do not care what you do in the meantime." I appreciated his sense of realism. He then asked me to participate in a program concerned with the total synthesis of steroid–cortical hormones for Royal Air Force (RAF) pilots. This work led to my first keen emotional involvement with strategic synthesis and its consequences.

The apparent necessity for our work arose from a rumor sent by the Polish underground that Luftwaffe pilots were being given such hormones, which are involved in shock conditions. I have never found out if this was true or if benefits would result, but the RAF had to assume both. Perhaps the rumor was in the same class as the one the British put out about fighter pilots being fed carrots to improve their night vision, to disguise the fact that the greatly increased night kills resulted from the still-secret radar. I was told recently by Lew Sarett that the same rumor initiated work in Merck & Co., even before the cortisone days, on the production of natural corticoids rather than analogs. The natural substances could not be extracted then in sufficient quantities for use, and partial synthesis was awaiting Tadeus Reichstein, Leopold Ruzicka, and particularly Russell Marker with his explorations of plants in Mexico. No natural steroid hormones had been made by total synthesis because of obvious formidable difficulties, so we had to assume that the Germans had found an active simplified analog.

Shortly before the War, following chance observations by Sir Charles Dodds, Robinson had synthesized diethylstilbestrol dimethyl ether (**20**). In 1938 I worked at the same bench as Leon Golberg, who had carried out the laboratory work. The story of DES is irrational because it results from fortunate circumstances, which I allude to elsewhere. It had no connection with a fancied resemblance of the skeleton to steroid nuclei (an a posteriori rationalization of Robinson). The high estrogenic activity of DES was most surprising and unpredictable. Many phenols were later found to have some estrogenic activity, which is rather structure-unspecific, in contrast to the very high structure-specificity of other steroidal androgenic, progestational, and cortical hormones.

The differences in biological activities of the hormonal series were then known, from chemical conversions of the natural series, to result from rather small structural alterations of specific groups at the strategic 3,11,17-positions and from the degree and type of unsaturation in ring A. Pertinent examples are the male hormone testosterone (**21**, R = Me, R′ = OH), one female hormone progesterone (**21**, R = Me, R′ = COMe), and cortisone (**22**). All of the highly active hormones, other than estrogens such as the phenolic estrone (**23**, R = H), have a cyclohexenone ring A structure. The overall shape of the nucleus is also important for biological activity, controlled largely by steric configurations at ring junctions.

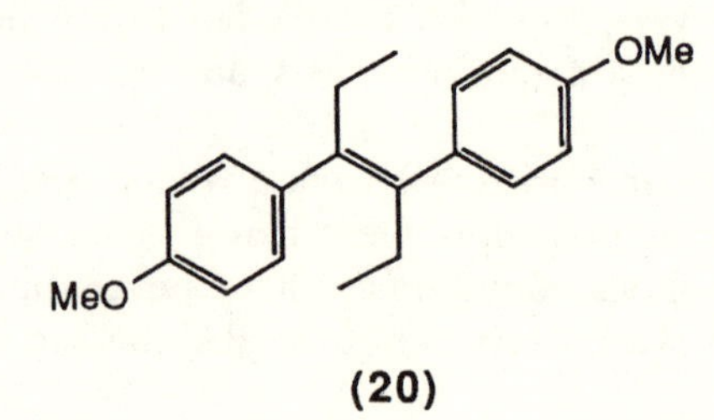

(20)

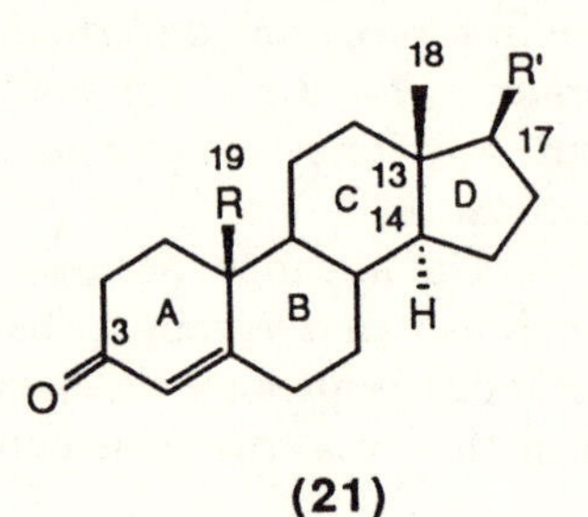

(21)

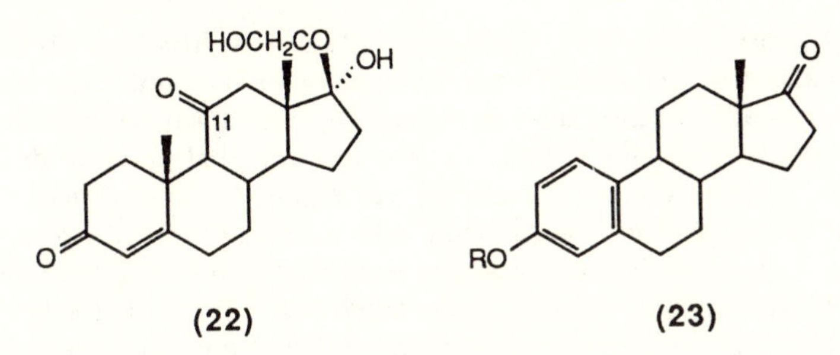

(22) (23)

No biologically active synthetic analogs of structure-specific hormones were then (1943) known,[38] with one doubtful exception. In 1936 Dirscherl[38] had reported the catalytic hydrogenation of the aromatic ring A of estrone (23, R = H) to form a ring A saturated product mixture, which was claimed to be a weak androgen. This mixture must have H in place of the normal angular 19-Me (cf. 21, R = Me); that is, it belongs to the 19-nor series (nor indicating lack of a C_1 structure). If it were possible, the use of aromatic material would avoid synthetic problems of stereochemistry and those concerned with ring reduction, as it did with estrogenic DES. We were forced to assume that such was most likely the path the Germans had taken. Accordingly, in 1941 Robinson asked me to attach the known oxygenated C_2 side chain, characteristic of corticoid biological activity, to aromatic nuclei of the biphenyl type, while Renée Jaeger did this for stilbestrol itself. All of the resulting aromatic analogs were, not surprisingly, estrogenic. They would merely have feminized the fighter pilots, causing them to grow breasts and producing sexual impotence (to the detriment of the RAF as a macho organization!). I had to think around the problem from new synthetic directions, but still with practical production much in mind.

An obvious rational approach I took on paper was to work away, structurally, from the natural series a step at a time, omitting one by one the molecular features (notably quaternary carbon atoms) that make total synthesis difficult. In this way, I believed that I might be able to define which of the possible biologically active analogs might be readily synthesized, while minimizing structural change.

Much later (1948) I became aware that M. Ehrenstein had claimed in 1944,[39] on very slender biological evidence, the progestational activity of a 19-norprogesterone with the unnatural steric configurations of the steroid ring system at both C-14 and C-17. This claim, even if true, did not establish any expected activity for the "natural" steric series, the total synthesis of which I had already embarked upon in 1943 on other grounds, and for which I ultimately provided the first and the only practical, synthetic method (Birch reduction), which Djerassi and Colton had to use, following my published intentions.

Djerassi claims the origin of his ideas only from the Ehrenstein clew,[40] and I do not doubt that he believes his account. But this episode raises acutely the question of relative credit for fundamental ideas and methods (typically British), against that for final practical development to accomplishment (American), no matter by what fraction of a whisker the application race is won. His contraceptive Pill was not the first used; that was Nilevar made by F. B. Colton (G. D. Searle) also by my method. History will justly remember him, but I hope for the right attributes. That judgement implies knowledge of the true overall historical situation.

My approach could be fitted into the total syntheses of simplified steroidal nuclei already carried out in Oxford, so initially I turned as good sources to the use of Robinson's **18** and its precursors. The substances might lead either to complete steroids or to appropriate structurally modified ones. I examined in parallel, at Robinson's suggestion but not very fruitfully, other routes to hydroaromatic nuclei related to **21**. In this effort I used variants of his annelation (the addition of enolate anions to cyclohexenones to yield polycyclic cyclohexenones with steroidal ring A types), which has the virtue of generating mostly the natural ring stereochemistry.

I started to think about the synthetic problems posed by the hormone system. The synthetic difficulties in constructing an alicyclic ring system could be traced to steric requirements, the achievement of which was not well understood in basic terms. As another major difficulty, the presence of quaternary carbons carrying the 18- and 19-Me limits the use of available aromatic precursors suggested by the six-membered rings. Before 1950, stereochemistry largely had to take care of itself, apart from a strategy of synthetic sequences involving enolizable carbonyl adjacent to a 6–6 ring junction (but no angular Me) to equilibrate this to the stable trans junction. Earlier steroid chemists, like I. N. Nazarov, virtually ignored such steric difficulties in their synthetic work, hoping for the best or simply conducting academic exercises. The catalytic hydrogenation of estrone by W. Dirscherl at least seemed to indicate that the 19-Me could be omitted with retention of some biological activity, despite stereochemistry (clearly mixed in that product).

Another synthetic problem I sought to solve was finding a facile way (other than the Robinson annelation) to make the ubiquitous α,β-cyclohexenone structure (e.g., of ring A in **21**), especially from an aromatic ring, by a means other than complete hydrogenation and reintroduction of unsaturation (which had steric drawbacks as well as being laborious).

In 1941 I tried some high-pressure hydrogenations of phenols with Raney nickel in alkaline aqueous solution. The results were interesting, including inhibition of OH-hydrogenolysis and sometimes direct formation of cyclohexanones. Resorcinol monoethers, so converted into methoxycyclohexanols, gave α,β-cyclohexenones by oxidation and loss of MeOH. None of this work was published because Robinson thought I should not be doing it, and he was my paymaster.

My cogitations on what was needed for analog synthesis (1941–1945) led to the experimental pursuit of the following topics:

- To improve the cyclization of the carboxylic acid **19**, the one bad experimental stage in formation of **18**;
- To achieve the stepwise partial reduction of aromatic systems,[41] preferably by generating eventually by a rather direct process, an α,β-unsaturated cyclohexenone structure in a steroidlike ring A, as characteristic of all of the biologically active hormones except estrogens (compare corticosterone (**22**) with estrone (**23**)). I wanted also to try such a reductive conversion within the stilbestrol skeleton into a testosterone analog (which I did later without biological benefit).
- To find ways to introduce, if absolutely necessary, the natural angular 18,19-Me groups missing from fully aromatic precursors. They could not be initially incorporated through structurally altered precursors of **20**.

I shall recount some of the practical results of my thinking, roughly in historical order, although most were obtained by parallel studies.

Polyphosphoric Acid: Cyclizations of Acids into Aryl Ketones

The discovery of how to cyclize **19** efficiently is a very good example of the role of accident and the perception of its consequences. Robinson had first brought about the closure of **19** by using phosphorus pentoxide in benzene, the only method to work. Some of his collaborators could obtain workable yields, others could not. The less careful ones, incredibly, showed the best results. Robinson himself found that traces of water were needed. Becoming impatient with the lack of progress, he went into the laboratory saying, "Of course it works; I have done it myself." He took a wet flask, impure benzene, and some decaying phosphorus pentoxide, and it did work. In the words of his collaborator, H. N. Rydon, in a lecture, "He flung the substances together in his usual inimitable fashion, and the reaction went." Robinson's public comment to this was, "Thank you, Dr. Rydon, for the bouquet you have thrown; I am trying to dodge the brickbat inside it."

In consequence, Robinson drew the correct conclusion; he dissolved **19** in phosphoric acid and then added P_2O_5. This procedure worked, but on a large scale the yields greatly diminished because of the great and uncontrollable heat evolved. I drew the further obvious conclusion,[42] dissolved the pentoxide first, and then carried out the cyclization at a controllable temperature. Renée Jaeger and I also first showed that this reagent (phosphorus pentoxide dissolved in phosphoric acid) is general for cyclizing in good yields arylbutyric and -propionic acids into cyclic aryl ketones.

Later the reagent was rediscovered independently under the name polyphosphoric acid by Snyder and Werber.[43] Their discovery was again accidental, resulting from using decaying phosphorus oxychloride (containing water) on phenylbutyric acids. They extended the method to solutions of phosphorus pentoxide in phosphoric acid ("polyphosphoric" acid, which indicates the importance of a name for indexing). Later still G. S. R. Subba Rao and I reverted to what was probably the Snyder reagent, $POCl_3$ in phosphoric acid, which is a more efficient cyclizing reagent than polyphosphoric acid itself, at least with carboxylic acids.

Introduction of Angular Methyl

In 1941, direct introduction of Me to form an angular structure on a quaternary carbon atom such as occurs in steroids could be mechanistically conceived, either by using a Me-cation generator attacking a correctly directed enolate anion (direct alkylation adjacent to carbonyl) or inversely through the action of a Me anion by nucleophilic attack at the β-position of a double bond conjugated with carbonyl (conjugate addition). In both cases an appropriate carbonyl-activated precursor structure is needed.

Cationic Methyl: Directed Enolization

When I began to work with the prototype reaction of an alkyl halide on an enolate anion, the problem was to ensure the formation of a quaternary carbon on the more hindered side of any carbonyl from which enolization can take two alternative directions (usually toward the less kinetically hindered side). The experimental outcome was then due to chance, depending on experimental conditions, with kinetic–thermodynamic factors in enolization and reaction not understood.

Because enolization involves removal of a proton, an obvious method of blocking enolization toward CH_2 adjacent to carbonyl is to remove both H atoms by reaction with some readily reversible blocking group. Robinson used the 16-piperonylidene derivative of 18-norequilenin methyl ether (**25**) to introduce an angular 18-Me (**26**), but he could not remove the blocking group.[42] At the time there was no method to determine the resulting unpredictable steroidal ring C–D stereochemistry, despite the availability of the same derivative of natural (trans C–D) equilenin methyl ether. Unlike the synthetic compound, it is optically active and could not then be directly compared. Advances in spectroscopy make this comparison routine now. Robinson dissolved my synthetic product and the natural one, separately, in sulfuric acid; one turned green, the other brown. He concluded from this primitive physical method that they were different. Some 6 months later, when I had remade my lost synthetic material, I showed that he was right.

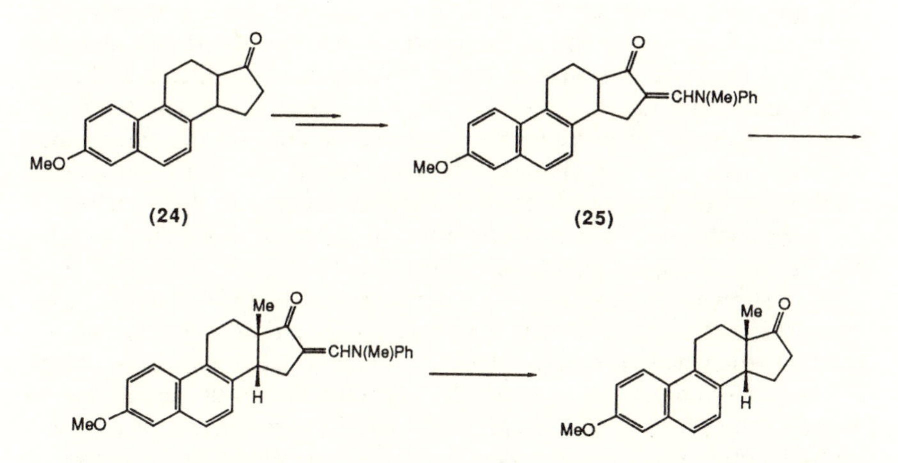

W. S. Johnson, who was working on the same problem about 1954, removed a similar 16-benzylidene blocking group by chlorination of the double bond. This procedure was inapplicable in Robinson's case because of concomitant nuclear aromatic chlorination. I therefore devised the methylanilinomethylene blocking group, which could be attached very simply in two stages to CH_2 adjacent to the carbonyl and be removed equally simply by hydrolysis.[41] We thus made isoequilenin methyl ether from **27** as shown. This product was the closest approximation to a natural hormone reached by the Oxford group before my 19-nortestosterone synthesis of 1948, described later.

I attempted, about 1946, to make directed enols by kinetic metal–ammonia reductions of α,β-unsaturated ketones, such as the octalones resulting from my reductions described later. However, I was not then able, with the methods available, to establish what happened on methylation. Gilbert Stork later succeeded. Subba Rao and I,[44] much later in a practically useless but personally satisfying sequence designed to exemplify the original synthetic concept, used a directed reductive enolization and methylation sequence, by then standard, to introduce an 18-Me in the correct steric configuration to complete the syntheses of equilenin and estrone, through the original cyclopentenone unsaturated Robinson acid precursor of **19**.

Anionic Methyl: Cuprous Salts and Grignard Reagents

Attaching a Me anion through an organometallic reagent (theoretically possible) to a double bond conjugated with a carbonyl is experimentally difficult. The carbon of the C=O group usually reacts preferentially, especially if the β-position is hindered, as is necessary if a quaternary carbon is to result. Because of my D.Phil. work on Grignard additions to alkylidenecyanoacetic esters, which did generate a quaternary carbon,[3] I was ready to perceive the synthetic significance of a chance but inspired observation of Kharasch.[45] This insight was that cuprous salts catalyze the addition of Me from a Grignard reagent to the β-carbon of the α,β-unsaturated ketone isophorone to generate not a ring-angular Me, but an analogous quaternary carbon.

I showed that this process could be extended to 2-octalones to introduce a true angular Me (**28**, R = H or Me) at a ring junction. But my products then had an unpredictable cis junction (also with a 6–5 ring), where a steroidal trans is needed. The random result caused me to abandon the process. In 1962 M. Smith and I improved the process as a high-yielding experimental procedure[46] and generalized the stereochemistry as axial addition, including to dienones. It took some 20 years after our original demonstration of the synthetic possibilities of this type of reaction before it received wide recognition. Then the recognition came largely through stoichiometric Li–Cu derivatives. This was the first of many synthetic applications. Because of Smith's tragic death, our work, apart from the basic clew, was not fully published.

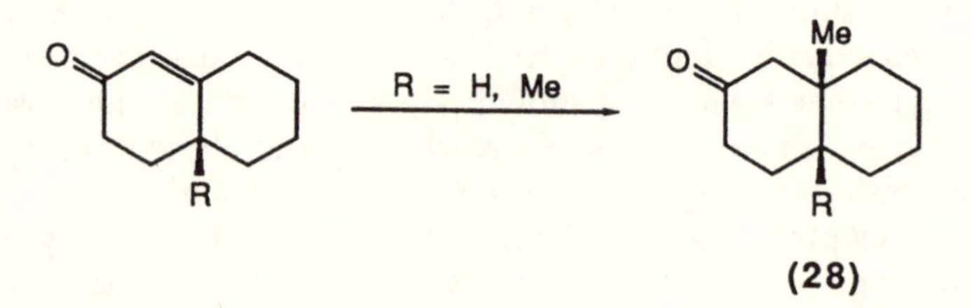

The mechanism was mysterious to me until much later (1960s) when I worked on transition metals (Research Set 4). Then it occurred to me that the preferred orbital interaction of the transition metal cation should be with less polar C=C, in contrast to a main-group Mg or Li cation with the polar C=O.

Methylene Addition: Regio- and Stereospecificity Dependent on an Unactivated Olefinic Bond

In 1963,[47] as the result of developing general synthetic reactions, we were able to use an unactivated C=C (between carbons 5 and 10 of our reduced nucleus) predictably to introduce stereospecifically an angular 19-Me. Our first example was achieved through dihalocarbene addition to the 5(10) double bond of the 3-ketal derived from the reduction product **29**, direct from estrone methyl ether 17-ketal following our reduction work.

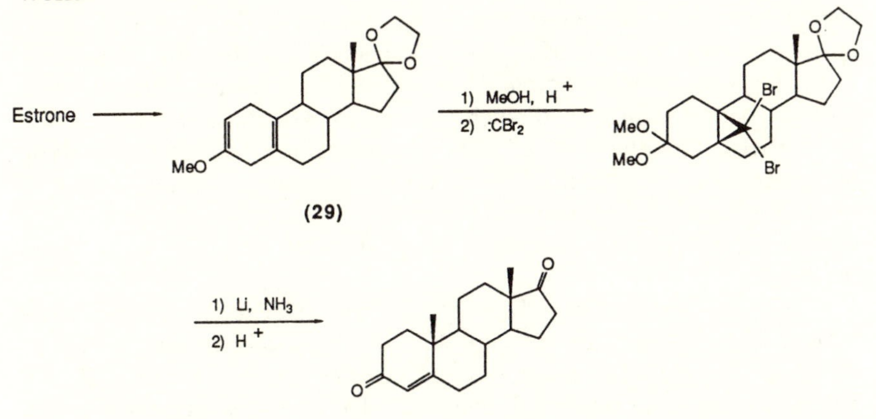

The overall sequence shown represents a direct efficient conversion of estrone into androstenedione. This total synthesis is still probably the easiest stereospecific sequence to nonaromatic steroids, given the efficient Herchel Smith–I. V. Torgov industrial total synthesis of the precursor, (+)-estrone. This most efficient synthetic approach to the general series has unaccountably been omitted from all reviews that I know of steroid total synthesis. With our other synthetic estrone stereoisomers, it also enabled us to make a number of stereoisomers of the testosterone type. I shall mention other methods we developed for forming quaternary carbons by kinetically alkylating mesomeric carbanions obtained from unsaturated ketones directly or by deconjugation processes.

All of these new procedures, and especially the general reduction methods to be described, illustrate that a novel reaction sought for one specific purpose usually proves to be of general and often unforeseen value. The partial reduction of aromatic rings was a direct result of our steroid-synthesis requirements, not incidentally applied to them later, as some authors prefer to assume. Focused strategic work, in my estimation, may therefore be more profitable than the uncommitted work so beloved of many research scientists.

Metal–Ammonia Reductions: 19-Norsteroid Hormones

In about 1941, when I was thinking about steroid synthesis, the methods for making specifically substituted polycyclic aromatic compounds were well developed through understanding of aromatic substitution reactions. The reduced cyclopentenophenanthrene nucleus of steroid hormones, containing mostly six-membered rings with one five-membered ring, could conceptually be made from aromatic precursors, providing certain outstanding problems could be solved. One problem was the partial reduction of such systems to provide functional groups as synthetic "handles". The methods existing in about 1941—catalytic hydrogenation, or reduction with sodium and amyl alcohol—could be used in a limited fashion to yield only fully reduced systems or systems containing at least one remaining intact aromatic ring (e.g., from naphthalenes).

Stereochemistry in reduced steroid ring systems was known to be biologically important for activity, but no theoretical concepts existed to direct specific configurational synthesis before Barton's crucial 1950 Nobel ideas on conformational analysis. The metal–ammonia reagent, which led to partial reduction of the aromatic rings, later provided an unexpected dividend in largely solving the steric problem for double-bond reduction as well, because of the formation of anionic intermediates in both types of reaction.[48]

The α,β-cyclohexenone ring A of steroid hormones (e.g., **21** and **22**) is known to be a key structure for biological activity,[49] so its presence in synthetic compounds is necessary. I felt that complete reduction of an aromatic ring and then reintroduction of unsaturation to form such a structure was obviously inefficient in a process leading to substances needed for real application. In addition, the approach was not in accord with my lazy temperament, which can be equated (charitably) with efficiency.

If precursors like **18**, **19**, or related substances were to be used, I believed that the reduction courses of substituted phenanthrenes, naphthalenes, and biphenyls must be defined in detail. Although some basic knowledge was available, few functional groups had been involved. But the main synthetic obstacle to our objective was clearly lack of a practical means to achieve the partial reduction of a monobenzene.

The first clue to resolving the situation was provided by the Cornforths, then still in Oxford.[50] They reduced 2-methoxynaphthalene with sodium and alcohol into its dihydro derivative (within the OMe-containing ring) and realized that this derivative contained an enol–ether group and was convertible by acidic hydrolysis into 2-tetralone (**30**). This key reaction later led to the Oxford steroid total synthesis. The initial clue was the reduction by A. Windaus of ring A of a 3-methoxy-ring A,B-naphthalenoid steroid, although he had not interpreted the overall mechanism correctly.

I thought then (about 1942) that my problem would largely be solved if this type of reduction could be brought about with a monobenzenoid system. The basis of the idea is illustrated by the simple model **31** for estrone methyl ether. If it could be reduced into **32** it should eventually yield the conjugated ketone **33**, the model for 19-nortestosterone (**21**, R = H, R′ = OH) and other functionalized 19-norsteroids.

The experimental difficulty was that sodium and alcohols cannot bring about such hydrogenations with monobenzenoid derivatives (except benzoic acids). Accordingly, I then (1942) surveyed the literature for any indications of partial reductions of monobenzenoid compounds and found clues in a new review[51] by B. K. Campbell and K. N. Campbell, for which I cannot be sufficiently grateful. I tried to repay this debt to the world later by publishing laborious reviews of my own. They reported that C. B. Wooster[52] had fortuitously found that benzene, toluene, or methoxybenzene can be reduced into dihydro derivatives by a combination of sodium and ethanol in liquid ammonia. This procedure contrasted with no reduction by sodium and ethanol, or by sodium alone in ammonia. Wooster was trying to measure the quantity of sodium remaining after reacting an excess of it with a heterocyclic compound that, being insoluble in ammonia, he had first sensibly dissolved in a layer of toluene. Workup with water or ethanol to estimate unreacted sodium by measuring hydrogen gas evolution surprised him, because there was none. He deduced that he had "lost" 2 atoms of hydrogen for every molecule of toluene present. He demonstrated that benzene gives 1,4-dihydrobenzene and that anisole also absorbs 2H per molecule.

He suggested that the toluene and anisole derivatives have the hydrogens in 1,4-positions; perhaps he just meant, correctly, that the double bonds were unconjugated. They are, crucially however, the 2,5-

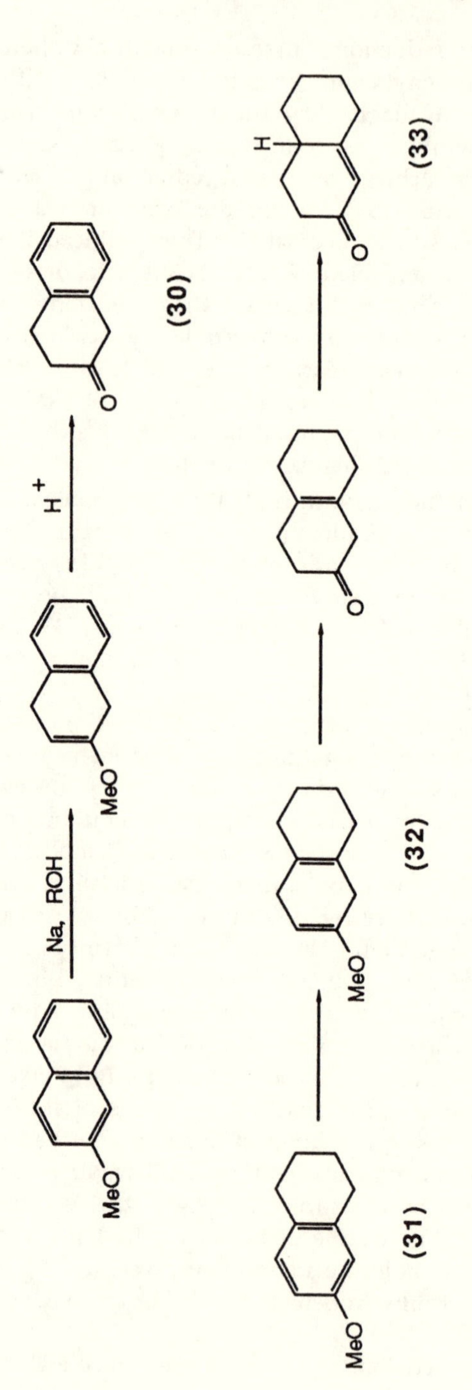
MeO
Na, ROH
MeO
H+
O
(30)
MeO
(31)
MeO
(32)
O
O
H
(33)

dihydro derivatives as I showed in 1943, thus making the anisole product an enol–ether, as in the Cornforth case, here of a cyclohexenone convertible into my desired type.

The critical factor for monobenzene-type reduction was the joint action of sodium and ethanol in the liquid ammonia. I first examined the reduction of methoxybenzene in January 1943, having previously achieved interesting results from naphthol reductions with sodium and *tert*-butanol in ammonia. I recall adding the reduction product of anisole directly to acidic 2,4-dinitrophenylhydrazine (used then to make ketonic derivatives). First a beautiful yellow crystalline derivative of the β,γ-unsaturated cyclohex-3-enone developed. Then, on heating, an equally beautiful orange-red crystalline derivative of the α,β-unsaturated cyclohex-2-enone appeared, proving that the initial formation of an unconjugated methoxydihydrobenzene was an enol–ether. I extended this in 1943 to the simplified estrone model **31**, which was reduced into **32** and converted by acid finally into **33**.

My initial results were published in 1944,[41] but without mention of the steroid connection for a security reason that ultimately proved illusory. The reaction was later named the Birch reduction by Carl Djerassi. He told me jokingly that this was to reduce the time and therefore the cost of telephone calls between Wayne State University (Detroit) and Syntex (Mexico City), where he was then alternately based. He was using my reduction process to make 19-norsteroids (to birch is now a verb). As Ralph Raphael later declaimed at a Gordon Conference, "The verb 'to birch' no longer means to thrash. No dictionary would be so rash as to put this meaning in first place." Later, he told me that the name was in my honor, which I accept. This nomenclature from about 1955 has led to the exaggerated rumor that I have been dead for years.

Development of the reaction was an emotionally satisfying precursor to some 40 years of work. I can remember the exact location at my bench where I stood when I made the beautiful dinitrophenylhydrazones. It was instantaneously obvious to me that this was the beginning of an era. Even the brilliant colors of the derivatives added to my exhilaration. The only other time I encountered such an intense burst of feeling was with the polyketide hypothesis. Both situations represented very sudden understandings. The lateral-control ideas of my organometallic chemistry (Research Set 4), although as personally satisfying in the end, did not arise so abruptly.

The method evidently could produce the type of α,β-unsaturated ketonic structure in ring A that was characteristic of the biologically active series,[49] but only an angular H could result because the starting material was aromatic. An extra step of then-unknown nature would be required to introduce an angular Me. Consequently, I wondered if a series of analogs could be active without having to insert this 19-Me; perhaps this was what the Germans had found. For example, is 19-nortestosterone (**21**, R = H, R′ = OH) biologically active? The precursor I needed to test the idea was obviously a derivative of aromatic estrone **23** to convert into ring-type **29**. But I then ran into grave political and practical difficulties. Robinson tried to divert me from pursuing the work (I was supposed to be making simple steroid analogs), and he would not supply me with estrone, which at 25 pounds sterling per gram would have cost me a fortnight's income.

Political Problems

For some Robinsonian reason I was technically an employee of ICI (Imperial Chemical Industries, Ltd.), through which a government grant was funneled. The only signed agreement I had with them was to submit work for consideration of patenting before I published it.

In 1943 I reduced, as models, a number of simple substituted aromatic compounds, wrote a paper, and sent it to ICI for examination before publication. I received in return a very rude letter asking what was I doing working in this field instead of making sex hormones. I was told that ICI had a cartel agreement with DuPont, to whom Wooster had assigned a patent. Under this agreement ICI was bound to stay out of this reduction field. Needless to say, I knew nothing of that and cared less. As an employee of ICI I could not publish, they said. I recall visiting Blackley, Manchester, to argue with them when much of Manchester was a smoking ruin from air raids. Harry Piggott, a chemist, supported me, but H. Payne, the research manager, was obdurate.

I took one of my first political steps by saying, "Do you or do you not want to patent before I publish? Inhibitory reasons concerned with cartel agreements would attract great public interest!" There had recently been a public fuss about ICI–IG Farben agreements still operating in South America, despite the war. Very ungraciously, Payne gave way. However, Robinson, who was their consultant, was reprimanded strongly for allowing me to do the work. He could hardly tell them he did not know what I was doing. Alan Johnson, who was then secretary of the ICI Consultants Committee, said that he told them I was a wild Australian with whom he could do nothing. The practical outcome was that when I took the paper to him to put his name on as usual, he said he did not want to hear another word on the topic. I interpreted this as permission to carry out independent research in the reduction area. That is why it is called the Birch reduction and not the Robinson reduction.

From left, W. B. Whalley, me, Vladimir Prelog, and Alan Johnson at an antibiotics conference in Dublin in 1959, where I first presented the idea of "partial biosynthesis" of antibiotics using mutants.

Looking back, I do not know why Robinson put up with me; after all, he organized to pay me. To do him justice, in a major 1947 BBC broadcast ("Science Survey", Ian Cox), he quoted my work as being one of the highlights of Oxford research in his time.

Robinson eventually provided me with 500 mg of estrone, not a large amount on the manipulative–analytical scale of the time. It was just enough to demonstrate the difficulty, because of insolubility, of reducing the Me ether of estradiol. My parallel model synthetic work on angular methylation was yielding results, and I widened the reductions to a basic study of mesomeric carbanions, as discussed later. With only one pair of hands, I was busy. Also, any steroid analog work belonged to Robinson, so it had a lower priority for me.

I first met Gilbert Stork in 1947, following the first London IUPAC Congress after the War. On learning of my problem, he said with characteristic generosity, "I have 5 g of estrone; I'll give it to you." This was enough estrone to make 19-nortestosterone (**21**, R = H, R′ = OH) via the precursor β,γ-unsaturated isomeric ketone,[53] briefly assisted by S. M. Mukherji in 1948–1949. I had chosen to make, instead of the more complex cortical one, the structurally simpler androgen with a 17-

OH group, to test any biological activity of the proposed new synthetic 19-nor nucleus within a structure-sensitive series. I envisaged in print the whole range of structural patterns characteristic of other hormonal activities, as I reported (1950) to my employers, the Royal Society. I do not know how this might have affected subsequent patents, because the matter was never tested.

After I had started in 1942 on my reduction norsteroid approach, an added impetus to seek the capability to make 19-norsteroids was provided by M. Ehrenstein,[54] who obtained by partial synthesis from digitoxigenin a mixture of almost certainly unnatural stereoisomers (C–D cis, 17-iso) of 19-norprogesterone (**21**, R = H, R′ = COMe). Nevertheless, on the basis of an exiguous test, he pronounced the mixture to be highly biologically active as a female progestational hormone. This clue (although never confirmed, so far as I know) started other workers, notably Djerassi, on the 19-nor trail.

In 1948 I sent my 19-nortestosterone and its β,γ-unsaturated isomer to R. A. Spinks at ICI for biological testing, but made the mistake of telling Robinson. He insisted on me withdrawing them, because he said he had an agreement with Charles Dodds of the Courtauld Institute (his collaborator on stilbestrol) that all Oxford compounds would be tested there. I heard nothing for about a year, and had impatiently published the synthesis[53] before I found out that the products are biologically active, the former as an androgen (about 20–30% of testosterone) and the latter as an estrogen.

Estrone had been made by total synthesis. According to the rules of the synthesis game, my product was the first totally synthetic highly active androgen (indeed, the first synthetic structure-specific steroid hormone). This work was published before the natural hormone syntheses of Woodward and of Cornforth and Robinson. The latter was also through a "relay sequence," not a continuous total synthesis, carried out via a natural steroid degradation product to which synthesis had led.

Significance of the Results

The important clue here was not that nortestosterone is less active as an androgen than testosterone, but that it has any pronounced activity at all. Well known from partial syntheses was that biological alterations resulting from structural alterations in one biological series are not necessarily paralleled in another. It has indeed full male-hormone anabolic activity, as I was told by Byron Riegel (Searle) in 1951. 19-Norcortisone eventually turned out to be inactive, at least in classical manifestations, and 19-norprogesterone was more active than the natural series, all unpredictable results. The oral activity of the progestational series as contraceptives was demonstrated by Gregory Pincus about 1953.

My work demonstrated for the first time a very marked hormonal activity of a highly structure-dependent type in an altered skeleton of normal stereochemistry. It provided a totally unique method to make the whole new 19-nor hormonal series.

Why No British Patents?

Todd gave me good support, including more estrone, when I moved to Cambridge in 1949. I wanted to patent the series and the method. Todd offered to approach the then-new (British) National Research and Development Corporation through Charles Dodds, who was apparently their consultant in such matters. Dodds refused his support, on the grounds that 19-nortestosterone is less active as an androgen than the natural hormone and would always be more expensive to make. The latter turns out not to be true, and the former was the very reason why it should be patented, considering that it has full anabolic activity. It and some derivatives are still used as anabolic agents, particularly (illegally) by female athletes because it has only a limited tendency to produce facial hair and a gruff voice. Carl Djerassi asserts that Dodds' reservations were correct. Nevertheless, I wanted to patent the unique synthetic method to make a whole novel series of analogs (which Carl, among others, later had to use). But with three young children, a dying mother, and still a very small income in 1950, I could not press the point. The whole development, including the later derivation of oral contraceptive pills, went to foreign countries, notably to the United States and Mexico.

The importance of my result is of course clearer in hindsight than it was in foresight, and I was very naive in those days outside my work. Dodds, my *bête noir*, owed his position largely to the fact that he had operated on King George. Because he was such a poor organic chemist, when he tried to make the propenylphenol anole by standard alkaline demethylation of anethole he accidentally obtained a dimer instead. It happened to be diethylstilbestrol (DES, **20**), the first powerful synthetic estrogen.

In Cambridge in 1950, my Ph.D. student Herchel Smith, then rather an experimental tyro, was trying to make 19-norprogesterone (**21**, R = H, R′ = COMe) by the standard method of adding acetylene at a steroidal 17-carbonyl position, to convert the resulting carbinol into the COMe progestational side chain. I then received a letter from Carl Djerassi (April 16, 1951) that said, "We have recently succeeded in preparing pure crystalline 19-norprogesterone by a modified Birch reduction of 3-methoxy-17-acetyl-1,3,5-estratriene .... My friend Gilbert Stork told me some time ago that you may have some biological results on your 19-nortestosterone. I am wondering whether the results will be published ...." He then offered to quote them. Unfortunately for Herchel and me, Carl had the required starting ring A aromatic progesterone on the shelf, made by industrial partial synthesis from diosgenin, whereas we had to make it with difficulty from estrone. I had written to Gilbert Stork in 1950 to inform him of our biological activities when he was a consultant of Syntex. The results were first mentioned publicly in the Annual Reports of the Chemical Society for 1950, submitted for publication on January 14, 1951.[55] In the Year Book of the Royal Society for 1951 (issued in January) I envisaged the whole hormonal series: "Almost all of the active hormones of the cyclopentenophenanthrene group, including testosterone, progesterone, and cortisone, contain such a cyclohexenone group (obtainable uniquely by my method), and the method thus provides a method of synthesizing analogs from aromatic starting materials."

The modification referred to by Djerassi was the use, pioneered by A. I. Wilds and N. A. Nelson (told to me in a personal communication), of Li instead of my Na. Anyhow, I had no Li available. The substitution worked well, particularly in the impure ammonia then available. Searle work on the isomeric β,γ-19-nor series had a similar synthetic origin. I recall a visit of their newly appointed research director, Byron Riegel, to me and Herchel Smith in Cambridge in early 1951 to discuss it.

Accounts of the history of this whole field and references are given in a symposium issue of the journal *Steroids,*[56] involving almost all of the survivors of the contraceptive pill and cortisone synthesis. The implications of that discussion supplement a number of assertions by Djerassi[57] on the origins of ideas in the field. Neither we, nor Djerassi had any thought at the time of an ultimate use as oral contraceptives; we were synthesizing possible sex hormones. The oral contraceptive activity (progestational agents had long been known to be active on injection) was discovered by G. Pincus about 1953. The work was done with reported progestational agents from the shelf, such as Djerassi's 17-ethinyl-19-nor precursor of the norprogesterone series and Colton's (Searle) β,γ-unsaturated isomer (5,10 double bond). Such oral activity was and is unpredictable.

An element of luck was the fact that the progestational series was first synthesized from the ring A aromatic series by my process. At first there was always a small proportion of the initial aromatic estrogen left in the product. When pure reduced compounds became available, it was found necessary to add some estrogen. If the 19-nor series had been made in some other way, time might have elapsed before this type of application was envisaged.

Herchel Smith continued for some time after 1956 in my department in Manchester, collaborating with me on sex-hormone synthesis as well as on biosynthesis. He wanted independence in the synthetic work, so I temporarily ceased my work in the steroid area. Having seen the personal consequences in Wisconsin of two rival steroid synthesis groups in the same laboratory, I did not want this in Manchester. Herchel took his work eventually to the firm of Wyeth (later American Home Products) in Philadelphia after I had introduced him to G. R. Fryers, who at that time wanted to set up a new laboratory for Wyeth in the United Kingdom. I persuaded Fryers that what Wyeth really needed was not to set up a rivalry with established firms on steroid partial synthesis from natural products, as they were considering, but to move ahead with total synthesis, and Herchel was the man to do it. All I gained was a dinner; Herchel did the rest himself.

Herchel's totally synthetic norgestrel (with an 18-Et instead of Me) and its analogs are probably the most widely used constituents of modern oral contraceptives. Much of the basic work on its synthesis was carried out in the Manchester University laboratories, as recorded in Ph.D. theses and in reports. I understand from Lord Dainton that the patentability of these basic results supported in the United Kingdom through Science Research Council scholarships was carried to Cabinet level. But they decided, following lessons learned from the development of penicillin, not to follow up on the matter of royalties. Herchel Smith later endowed several Chairs at British universities (UMIST and Cambridge).

I did not personally obtain further biologically applicable results, although we achieved a desirably high anabolic:androgenic ratio in a D-homo-19-nortestosterone.[5] Some interesting biological activities were found with ring-A tropone and tropolone hormones.[58] The discovery was initially made through a personal whim because ring A in these is partly aromatic in character (estrogen) and partly unsaturated ketone (androgen). I thought it would be interesting to see which sex won (resulting in schizophrenia?).

I have given an historical perspective of the field in greater detail,[56] and Djerassi also has recorded the story from his totally personal viewpoint.[40,57]

The Synthetic Importance of Mechanistic Studies

Understanding reaction mechanisms is not only an objective in itself. It is essential as a base for synthetic applications. Robinson's predictions of reaction products in the 1920–1930s, impressive because they were mysterious in origin to many of his fellow chemical cooks, were based on his understanding of the electronic nature of bonding changes in reactions.

It is typical that a fundamental understanding of mechanisms acts as a "creation center" from which radiate other discoveries and applications, mostly quite unexpected. Developments can arise because mechanisms suggest new applications, or they can make newly possible the fulfillment of foreseen requirements. A complex web of discovery–invention results, often leading to major new radiating research centers.

I illustrate briefly here a set of webs that grew out of my original strategic partial reduction of a benzene ring. The results have been important for me personally in setting up many creation centers: for the theory of electron and proton additions in reductions; for reactions of mesomeric carbanions; and for the distinction between kinetic and thermodynamic control of product formation. They also led to new sources of unique materials for novel synthetic processes through some new reactions developed by yet more ancillary mechanistic ideas and philosophies.

A Fundamental Type of Process

Reductions represent one of the most fundamental of organic reactions: the addition of solvated electrons to molecules, and the subsequent reactions of the resulting mesomeric radical anions and anions. A spectrum of reagents is experimentally available in the metal–ammonia series (variable acidity of proton source from ammonia upward, different metals, temperatures between –30 and –80 °C, mixed solvents, and rationally altered substrates) that make the processes uniquely adaptable for studying mechanisms and for producing in practice a controllable range of products. I cannot go into fine details here, but there are some reviews[59] on the subject.

In 1980,[60] in collaboration with Leo Radom (I asked the questions and he gave the theoretical answers), the four possible successive stages of reduction of a substituted monobenzenoid compound were elucidated theoretically by quantitative ab initio calculations. The conclusions agree with our experimental findings and earlier more intuitive theoretical conclusions. For example, they explain why methoxybenzene is reducible faster than benzene and why unconjugated dienes are the overall (kinetic) result.

We had explained[61,62] the key requirement of the presence of a proton source[41] for consummation of reduction with compounds like benzene by an initial electron-addition equilibrium with a radical anion of type **34** far on the side of dissociation. The predictably low basicity of radical anions does not permit their protonation by ammonia alone. Nevertheless, they can be protonated by the more acidic alcohol, with continuous reestablishment of the electron-addition equilibrium involving the benzenoid precursor. The sequence can thus finally proceed by protonation to complete reduction as shown via **35**, leading to the neutral kinetically-determined unconjugated dihydro derivative of type **36**. Hydrogen gas is not normally competitively evolved because a hydrogen atom is a strong acid in liquid ammonia and remains dissociated into solvated ions (metal cations and electrons) in the absence of catalysts such as the transition metal Fe. I showed[41] that Fe prevents reduction by catalyzing competitive removal of the hydrogen as H_2 presumably through combination of two unsolvated hydrogen atoms. G. S. R. Subba Rao later adapted this reaction to develop a very useful method of limiting reductions to readily reducible polycyclic aromatic compounds.

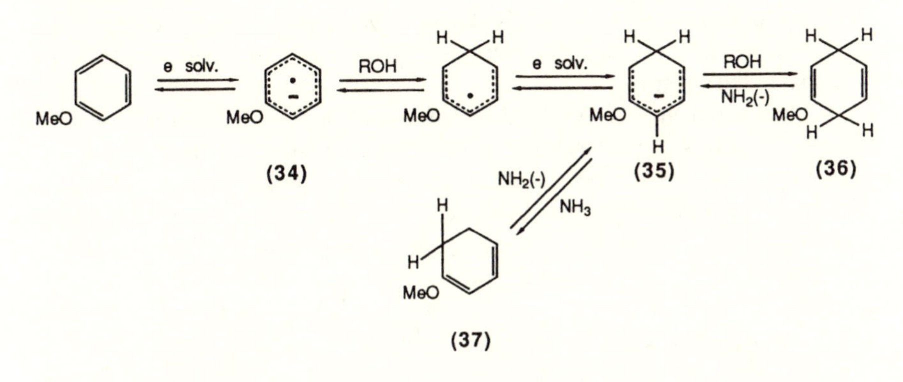

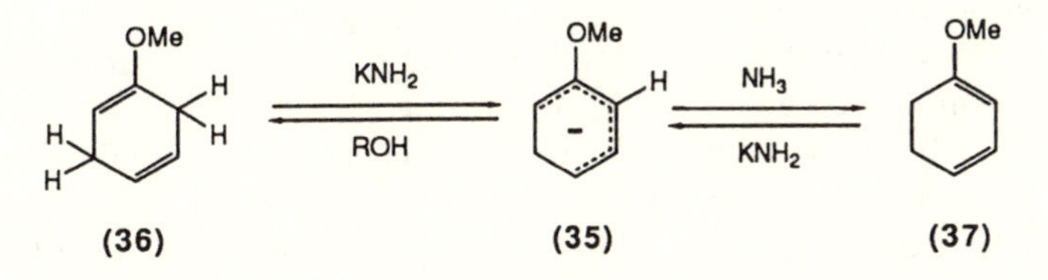

We also needed to rationalize the positions of addition of protons to the intermediate anionic systems, relative to various types of substituents. We started by considering the structure of the products. The first step is the addition to type **34** of a proton either ortho- or meta- (as shown here) to an OMe or alkyl substituent, to yield type **35** by further irreversible electron addition. I have always instinctively favored the initial meta- situation, but the structures of final products with two added H atoms indicate that either or both situations may be involved. Reaction of the mesomeric **35** with a second proton under kinetic control to the central position of the anion forms unconjugated **36**.

Kinetic Versus Thermodynamic Control

Examination of this protonation step under reversible and irreversible conditions is amenable to experiment. I deal with it here because of its general importance in that it first established firmly, on both a theoretical and an experimental basis, the general theory of the distinction between products whose structures are determined by a reaction rate or an equilibrium position.

Asking why the reductions yield unconjugated cyclohexadienes is indeed equivalent to asking why they give dienes at all, because a conjugated alicyclic diene product (like **37**) not stabilized by aromaticity would be reducible more readily in situ than its benzenoid precursor. Thus reductions under strongly basic conditions, if they occur at all, give the further tetrahydro reduction products (here cyclohexenes or methoxycyclohexenes).

Examination in 1947–1950 of equilibria in the OMe–cyclohexadiene series[63] gave me the experimental answer. A mesomeric anion of type **35** can act, according to experimental conditions, either as a turntable (reversible protonation) to yield the more stable conjugated isomer of type **37** or by kinetic protonation fully into the less-stable unconjugated diene **36**. The anion of type **35** can conveniently be experimentally generated directly with a base like KNH_2–NH_3 not only from the unconjugated **36**, but also similarly from the conjugated diene of type **37**. As I showed, **37** can be experimentally converted by kinetic protonation (or alkylation) into an unconjugated isomer like **36**. This sequence demonstrated in principle for the first time a route for making a less thermodynamically stable unsaturated isomer from a more stable one. The whole affair arose inevitably from trying to explain the structural nature of the metal–ammonia diene-reduction products. I rapidly realized that the reductions embody two of the most fundamental of all organic reactions: the addition of electrons to unsaturated molecules and the protonation of mesomeric anions. Both reactions occur under uniquely controllable circumstances with these reagents.

Deconjugations

I employed my new idea in a number of ways, notably for the first kinetic deconjugation of α,β-unsaturated ketones into the β,γ-isomers by kinetic protonation of the appropriate fully conjugated mesomeric extended (stable) dienolate anion. For example, in 1950 we carried out,[63] as a demonstration of the theoretical idea, the final stage of a total synthesis of cholesterol. Cholest-4-enone **38** (the Cornforth–Robinson total synthetic product in this area) was converted into **39**, which was known to be reducible by lithium aluminum hydride (LAH) into cholesterol. Robinson thought that somebody would find out how to do this efficiently (it had been done formally), although Cornforth wanted to take a clearer ring-A approach to cholesterol itself. This difference of opinion held up their synthesis for some time.

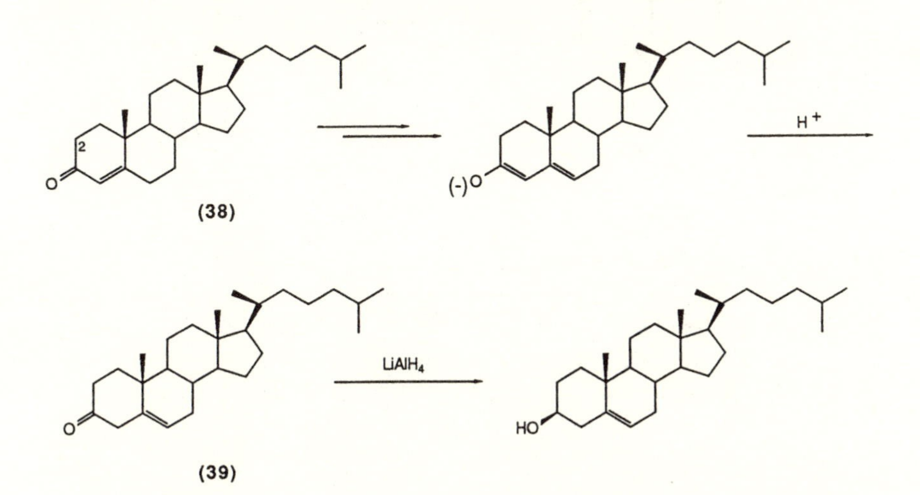

Also, because carbanions can be reacted with other electrophilic reagents such as alkyl halides, we devised a number of related processes. One such process was the conversion of **40**, which we obtained by modified reduction procedures, by deconjugation into the Miescher-ketone, the Cornforth and Robinson steroid starting material **41**[64] (they made it differently). We used the first reductive alkylations of benzoic acids to produce quaternary carbon atoms (e.g., **42** into **43**) and the similar kinetic alkylative deconjugation of α,β-unsaturated acids (e.g., **44** into **45**).

It took some 30 years for such reactions to be recognized as part of the synthetic repertoire. An example would be the alkylation of the cyclohexadienyl carbanions, which we soon showed to be efficient alternatives to dihydroresorcinols for synthetic purposes (e.g., for a projected steroid synthesis).[64] This neglect may have initially resulted from dislike of my base, KNH_2–NH_3 (not at all terrifying except in appearance) or from my incorrect assumption that most people can see the obvious without its expression in words of one syllable. General methods clearly require exemplification with real applications, for which I did not have the support.

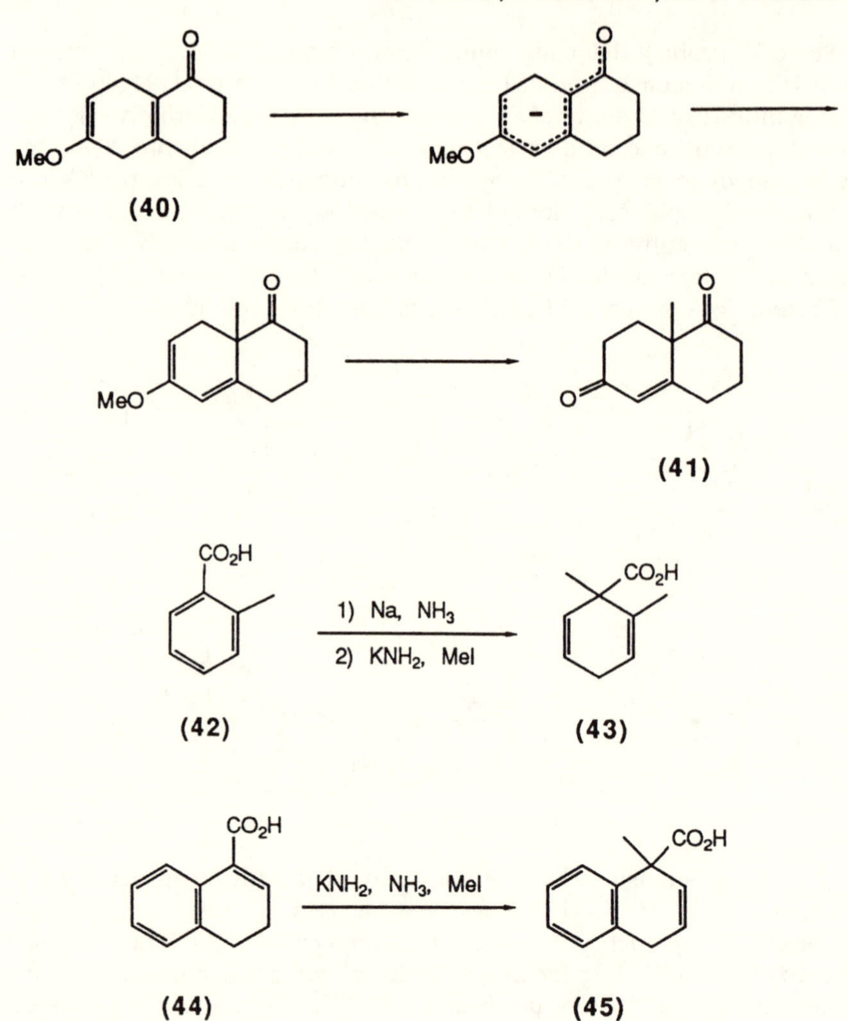

Further Synthetic Uses of Substituted Dihydrobenzenes: Some Overall Philosophies of Synthesis

The unique availability of dihydrobenzene derivatives with unsaturation regiospecific to substituents, especially oriented enol–ether groups not available by any other process, was for many years a creation center for me, almost a monomania, when I was faced with any synthetic problem. It was a highly directed personal approach, but a successful one, which would have been more so if I had been able to marshall sufficient logistic support.

Many applications depended on the high and selective reactivity of the enol–ether bond. This bond can be selectively reacted directly with carbenes, yielding the first OR-substituted cyclopropanes.[65] The result is extremely facile syntheses of tropones and other enlarged rings, such as the steroidal tropone, part structure **46**, eight-membered rings like **47** or their collapsed 6–4-ring products, or alternative regiospecific steroidal 2-Me introduction as shown in **48**. The high enol–ether reactivity can also be used to cover that double bond as a ketal. Any remaining unsaturation can then be reacted (e.g., with a carbene) as in the steroid angular methylation already mentioned from **29**. Halocarbene adducts of reduction products in the polycyclic aromatic series have been extensively employed by E. Vogel to form many bridged macrocyclic aromatic compounds.

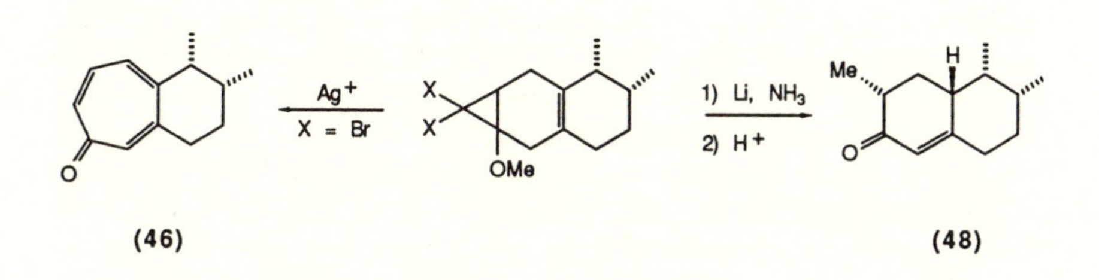

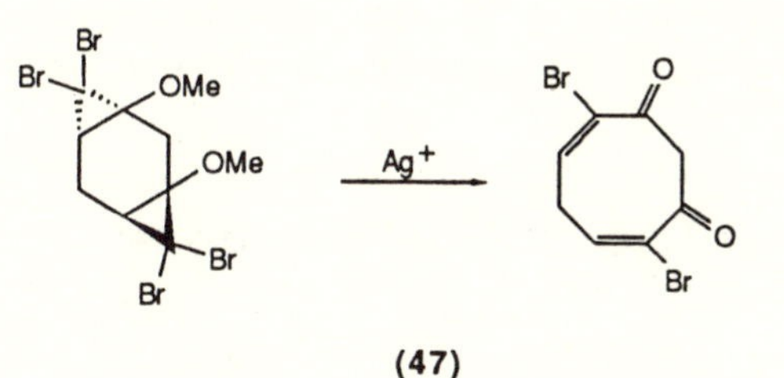

Syntheses Based on Availability of Unique Dienes

The ready availability, following the Birch reduction, of a further major series of specifically substituted conjugated cyclohexadiene systems was the key to another set of applications through the reactions of Diels (with Alder or with Rickert). This supply permitted me to exemplify in practice some novel concepts in the overall philosophy of synthesis, about which I started to think in the early 1950s, going beyond my previous classical obsessions.

Strategies

Modern strategic "retrosynthetic analysis" of a structure is simply a new name and symbolism for the basic concept that has been used ever since the synthesis of indigo in the 19th Century. It is now guided by an increasing knowledge of synthetic methods for bond formation. A new name is often confused with a new concept, to the benefit of its inventor, who then claims ownership. That is not true here: E. J. Corey needs no such artificial support of reputation.

In planning the strategy of general synthesis, I came up in 1953[66] with the concept later styled "convergent synthesis". I quote: "... a synthesis will be more efficient if two approximately equal halves are built up and joined together, rather than by successive additions of small pieces to a core. The convenience of the work is increased because more manageable amounts of starting materials are involved and because losses occur a lesser number of stages from available materials. Furthermore, in many cases if this procedure were possible it would utilize more efficiently the different elements of the desired structure which happen to be present in available molecules."

The chief difficulties with implementation of this idea were to attain the necessary selective reactivities for specific junctions of usually polyfunctional molecules and to achieve steric configurations. My desire thus to find stereospecific ways to connect regiospecifically major molecular pieces later affected my thinking on the strategic aspects of the organometallic work in Research Set 4, which provide one example of superimposing highly selective (including enantiomeric) reactivity.

Experimental Use of Diels' Reactions

The generalized Diels–Alder reaction seemed to me to be a good method for joining major molecular "slabs", following ideas I had gained from my literature review of it in 1950.[59] The reaction forms new C–C bonds with the minimum of functional group requirement: a conjugated diene in a purely carbon system, together with electron-accepting minimal activation of unsaturation in the other reagent. Usually no other drastic reagents are required, only the right thermal conditions, so a number of reactive functional groups can be present in the constituent molecules without detriment to yields. This fact greatly reduces the need for "protection" procedures and selective reactivities that are involved in reactions with standard polar groups and leads to abbreviated reaction sequences. Steric results are uniquely predictable, as I shall mention later.

An apparent synthetic drawback to the Diels–Alder-type reaction is the fact that two new C–C bonds are necessarily formed. If the diene is cyclic, as it would be in connection with my obsession with products from aromatic Birch reductions, the addend must contain a bridged ring, which is often not synthetically desirable. Production of a new nonbridged system therefore will require the subsequent fission of one of the new rings.

I exemplify my conceptual solutions to synthetic requirements for nonbridged rings using D–A processes, based on the novel unique availability of specific, structurally predictable, multisubstituted conjugated 1-methoxycyclohexa-1,3-dienes. These compounds are obtained by conjugation of the Birch-reduction unconjugated precursors either to an equilibrium mixture (as **36** into **37**) or, with more overall practical efficiency, by direct use of unconjugated reduction products (such as **36**) with continuous in situ conjugation during the addition process. This conjugation (in the OMe series) can be catalyzed by the dienophiles themselves or, more efficiently, by addition of a structurally related but nonadditive charge-transfer catalyst such as dichloromaleic anhydride. Addition products are thus obtained in higher than the "equilibrium conjugated-diene" yield.

Because of the specific electronic orientation of addition of the dienophile relative to the 1-OMe on a diene (e.g., in forming **49**) and because of regiospecificity of OMe–diene addition to unsymmetrical hydroxylated quinone dienophiles (due in that case to C=O polarization caused by hydrogen bonding).[67] Fortunately, therefore, a new set of unique opportunities exists for fully specific synthetic attainments. This finding applies particularly to making the novel aromatic substitution patterns of my polyketide–biogenetic natural products (Research Set 3). By contrast, standard aromatic substitutions yield this structural series (e.g., with meta-OR) only with great difficulty.

Diels–Alder Addition Followed by One C–C Fission: Regio- and Stereospecificity

How is a new bridged ring to be opened in an adduct of my cyclic dienes? Standard methods could be envisaged, such as cleavage of the double bond inevitably present. However, a much more attractive general process was provided by a fortunate observation. My very able collaborator Doug Butler, in examining a quinone adduct of a 1-OMe-cyclohexa-1,3-diene, found that the more frequently it was crystallized the less pure it became, and that eventually it had turned into something else.[68] He correctly interpreted it as the result of an acid-catalyzed fission of that one new bond formed, which terminates at the carbon carrying OMe. Like many reactions, this is predictable afterward as a kind of retro-aldol fission. It was extended to a new general type of synthesis[69] of a wide variety of functionalized 4- and 4,4-substituted cyclohex-2-ones. For example, **49** and **50** are transformed as shown, with an otherwise virtually unobtainable induction of stereospecific quaternary carbon atoms in a final nonbridged system.

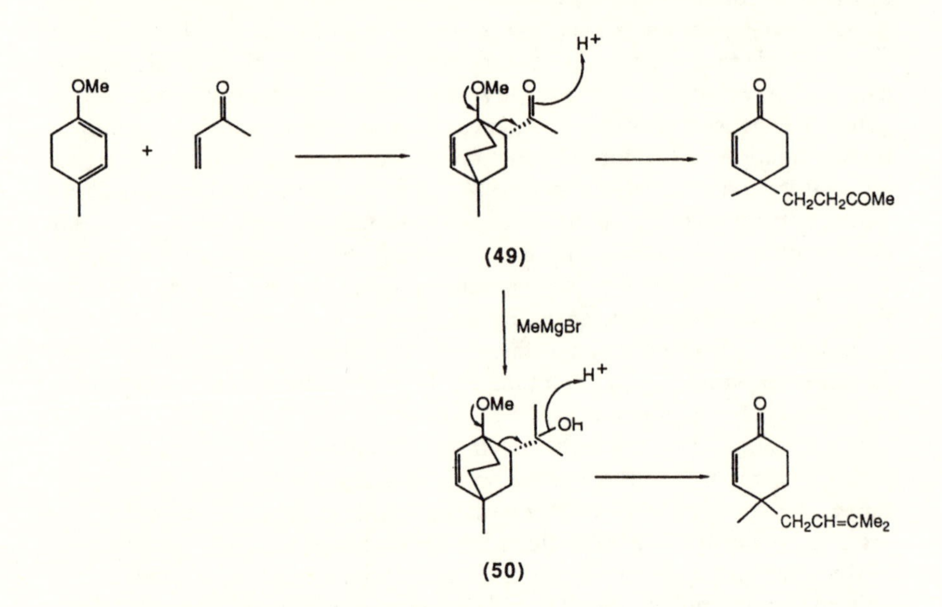

Diels–Alder reaction rates are sensitive to the substitution of the dienophile but not particularly to that of the diene, so production of many required quaternary centers often presents no experimental difficulties. The types of structure resulting, in connection with new transformation procedures (e.g., based on **49** and incidentally incorporating a terpene unit) form a wide range of precursors for further synthesis.

The steric consequence is largely predictable through the known rational course of the lateral molecular approach in Diels–Alder additions. The overall regiospecific orientation of the addenda is determined in the first instance by the electronic effects of substituents such as OMe and COR. In the case of the 1-OMe dienes, which were my principal interest, this group is found in products to be adjacent to the COR (e.g., **49**), a result that can be theoretically rationalized. The detailed stereochemistry of the product results from the requirements of the transition state. The initial configuration of the substituents is retained on the double bond of the dienophile, whereas flexible substituents are positioned to avoid overall crowding as the addenda forms the two new C–C bonds simultaneously. The overall synthetic importance of the combination of reactions is that one of the two new centers may thus be formed permanently in an orientation that is difficult to reach if it is made singly by any standard reaction. The following two rather different types of application illustrate this concept in convergent syntheses.

Juvabione. The only stereoselective synthesis of the insect hormone juvabione[70] (**53**) depends on the possibility of physical separation of the two stereoisomeric bridged-ring components (**51** and **52**, R = CH_2CHMe_2) formed together in the Diels–Alder addition. These components have differing physical properties, arising from forced proximity of functional groups, which can be used to separate them. The overall process leads after ring opening in each case to the pure monocyclic distereoisomeric compounds.

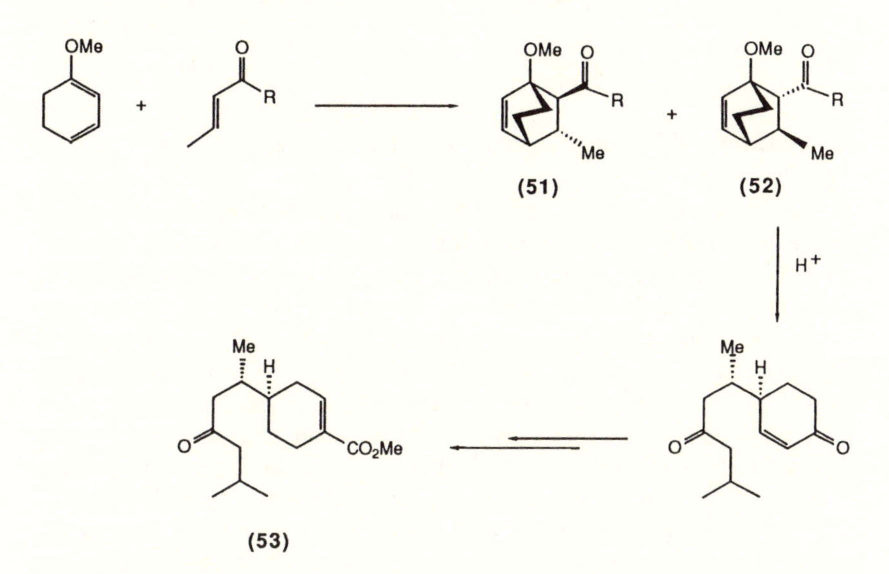

The monocyclic final products themselves, of type **53**, cannot be efficiently separated because the "floppy" side chains result in almost identical physical properties. This type of approach is stereoselective rather than necessarily stereospecific (which depends on the orientation of the initial addition, here not as favorable as I had predicted on the basis of models with R = Me).

Nootkatone. The stereospecific formation of the skeleton of the perfume and flavor component nootkatone (**54**), obtained through a joint effort with my collaborator K. P. Dastur as shown,[71] results in final production, after ring-opening, of the more crowded remaining nuclear Me center with the *cis*-di-Me, resulting from the first less-crowded dual-center transition state. This steric synthetic result is difficult to obtain by any standard procedures that directly introduce a single Me.

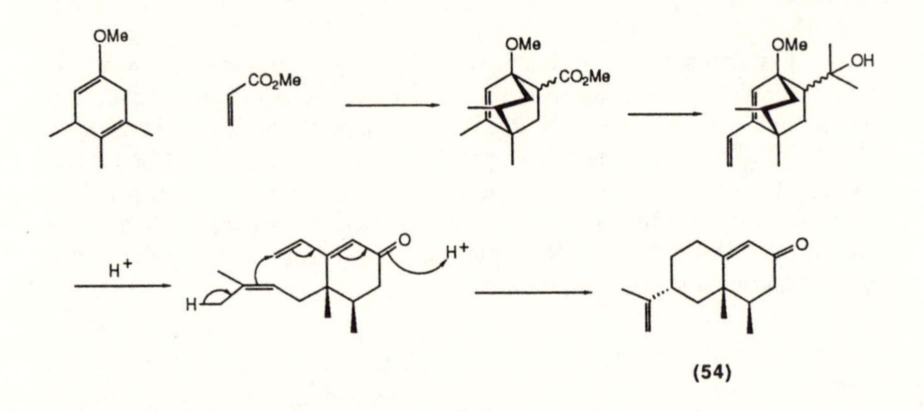

Diels–Alder Addition, Followed by Two C–C Fissions

The known Alder–Rickert reaction, usually carried out by reaction of acetylenes with cyclohexadienes, forms aromatic products by breaking two C–C bonds of a bridge in the resulting bridged cyclohexa-1,4-diene, but clearly not by breaking the new bonds formed in the initial reaction. I had previously used[72] this process in standard fashion to determine the structures of the conjugated dienes obtainable by conjugative isomerization of my substituted benzene 1,4-reduction products.

I was struck by the polyketide substitution pattern of the aromatic products thus obtained, such as methoxyphthalic ester (**55**) and later the methyl ether of my key biosynthetic polyketide metabolite, methylsalicylic acid, which I discuss in Research Set 3. We were able in this fashion to synthesize polyketide naphtha- and anthraquinones[67] and also, for the first time and efficiently, the medically and biosynthetically important mold product mycophenolic acid (**57**), our route to which is shown schematically.[73] Recently my former collaborator Subba Rao[74] ingeniously used the approach to synthesize a number of polyketide mold products: the dimethyl ether (**14**, R = Me) of our earlier medium-ring lactone curvularin and other biologically important metabolites (e.g., lasiodiplodin).

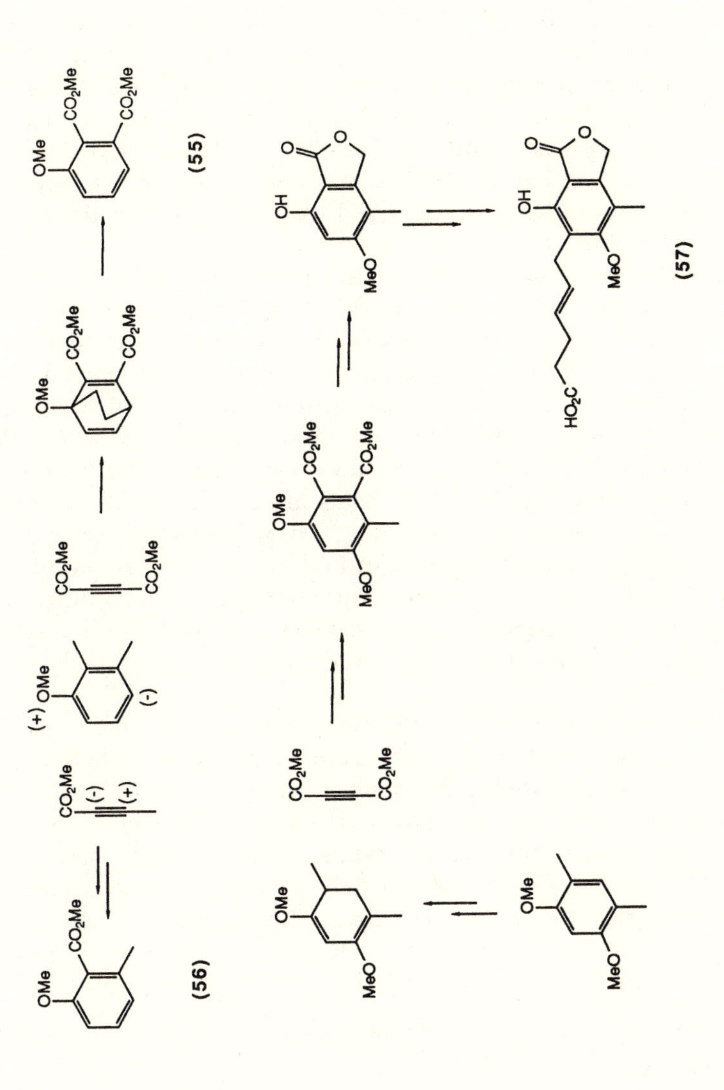
OMe
CO2Me
(55)
(56)
(57)
(+)
(-)
MeO
OH
O
HO2C

Other Reduction Fields

Metal–Ammonia-Induced Fission Reactions. Because of functional groups like OH and OR in substrate aromatic molecules, the reductions of which I was systematically examining, I coincidentally encountered a number of metal–ammonia-induced C–O or C–N(+) fission reactions. These obeyed useful structure-specific rules, for example with certain alkyl-substituted allyl alcohols.[75] The dealkylation of substituted phenol ethers (in absence of a proton source) has specificities different from standard acid-catalyzed fissions and is sometimes uniquely useful in synthesis.

Methylenedioxy groups are split in only one sense in relation to substitution, unusually with removal of one oxygen from an aromatic ring. Practical uses include alkaloid degradations. Such fission reactions turned out to have applications in structure work (Research Set 1) and in synthesis. The theory also includes how to avoid fissions if necessary. Some facile syntheses of unusual aromatic substitution patterns resulted. I cannot summarize these syntheses briefly here (cf., e.g., reference 62), but note that the results are typical of the surprises encountered in systematically examining a broad field with an open mind about what is relevant and rejecting nothing as irrelevant if the unpredicted happens. My approach to observed reactions differed from that in many investigations undertaken by others for specific synthetic objectives, where the unexpected was often rejected.

Heterocycles. I have also reviewed, in theory and practice, the very large field of the metal–ammonia reductions of heterocycles,[76] where theoretically predictable unique synthetic capabilities now result in the formation of partially hydrogenated systems or in unique anions for further conversions. These results are theoretically based on the ability of heteroatoms to potentiate the acceptance of electrons and to stabilize negative charges under definable experimental conditions.

Developing Activities Related to My Official Situations

To a small extent in Sydney and to a much greater extent in Manchester and Canberra, I was required to accept many obligations (sometimes accompanied by privileges) that affected my science either positively or negatively. Scientists who have made public progress in technical areas are consequently deemed able to perform in others. The ability to cope

Bob Woodward at Harvard in the mid 1940s. (Photograph courtesy of Harvard University Archives.)

Gilbert Stork (on the left) and Derek Barton, about 25 years ago. (Photograph courtesy of Carl Djerassi.)

successfully with pragmatic chemical research appears to generate characteristics that are applicable in science–technology and managerial organization as well, if one has the confidence to apply them and the ability to appreciate all of the factors involved, including the human ones.

From left to right: Carl Djerassi, Michail Shemyakin, Derek Barton, Bob Woodward, Alexander Todd, Vladimir Prelog, and Melvin Calvin at the IUPAC Meeting, Zurich 1955. I have talked about most of these eminent scientists and the roles they played in my life in this autobiography. Djerassi, Barton, Prelog, and Calvin have all contributed their autobiographies in this series.

Become a Professor and See the World

Commencing in Manchester, I had to accept the obligation to travel frequently to scientific conferences and to industrial connections that added to my sense of scientific realism and even a little to my income. My pejorative student-generated title, "BOAC Professor" (after the British Overseas Airways Corporation), was libelous because my absences were very brief. I was away fairly often for a few days or weeks; those who went for a year were not missed. I never did so, and I often took no vacations.

Among other visits, I spent two periods in Nigeria, under Nuffield and Rockefeller grants, helping to set up natural-products research in Ibadan and elsewhere. This experience provided a fascinating insight into the Third World, which was very useful to me in later UN program developments. Manchester University, being self-assessed at the center of the scientific world, did not provide study leave and therefore did not pay my salary. However, the university graciously continued my superannuation, provided I paid my share of it. I thought later that Australians did not realize how fortunate they were, and even still are, in having officially recognized leave.

The student Chemical Society event of the year was the showing of the 8-mm films collected on my travels, like the Ramadan festival in Nigeria (with High-Life gramophone accompaniment) or the Carnaval in Quebec, with the songs of the voyageurs, ice sculptures, and general *joie de vivre*. My talks included miscellaneous jokes, both scientific and political. For example, here are some Soviet bitter jokes.

- Customer in empty shop: "No meat?" "I'm very sorry, Madam, this is the no-fish shop; the no-meat shop is across the road."

A 1989 caricature of me as a lecturer, done by an American visiting Australia, whom I met by chance.

- From the mythical "Radio Yerevan": "What is the difference between capitalism and socialism?" "Under capitalism people exploit people, whereas under socialism it is entirely the other way around."
- "What does a Soviet symphony orchestra resemble on return from abroad?" "A quartet."

- The dilemma of the absent-minded shopper outside the Gastronom with an empty basket: trying to decide whether he had just come out, or not yet gone in.

People used to say, "How lucky you are to travel so much!" After about 2.5 million miles, I tend not to agree. Being in places is often pleasant, as distinct from getting there or leaving. On several occasions my life was genuinely in danger, but I can tell some amusing stories, mostly concerned with the socialist world.

Before going to Romania to negotiate a treaty on behalf of the Royal Society, I took the required documentation to their Embassy in Kensington Palace Gardens. They said, "No problem, just let us have your passport for a month." "But I cannot; I need it to go to Switzerland. Cannot I arrange a date and a time?" So I had a letter, agreeing to let me have a visa on a Monday at 3 p.m. (I needed to leave for Bucharest on the following Tuesday.) I turned up on schedule, only to be met with puzzled stares. I produced the letter. "He is in Edinburgh; we know nothing about it. In any case, only the ambassador can sign it, and we don't know where he is. Come back at 9 a.m. tomorrow." "But I cannot, my plane takes off at 8:15." Shrugs and smiles: "That's your problem!" So I made individual throat-cutting gestures: "This is a mission at government level. Somebody's head will roll. I don't say it will be yours (gesture) or yours (gesture) or yours (gesture), but somebody will be executed!" The smiles faded. "What hotel are you in? Give us your passport and we shall see what we can do." At 11 p.m. I went to bed; at 1 a.m. there was a knock-knock on the door. A man handed me my passport and visa. He had waited on the ambassador's doorstep until he returned home.

Other comic-opera scenes occurred in Romania. For example, with people sleeping in the corridor outside my hotel room, my bag sat in solitary splendor within, while I traveled for a week. I wanted them to store it and use the room. "But the Academy has allocated the room to you; we can do nothing else with it."

As I left Romania there was another "socialist" episode. My companions were Nicholas Kurti and Arnold Burgen. On Saturday we were told, "I am afraid your plane for Sunday morning has been cancelled; it has to fly a trade mission to Moscow." "Well, when is the next one?" "Tuesday afternoon." We all had teaching commitments on Monday in England, so we began to thump the table. In particular Kurti, of Hungarian origin, could express himself vehemently in a number of languages, including by this time a variety of choice Romanian phrases. Finally they produced a large Ilyushin Tarom plane at 5 a.m. to fly the three of us to Swechat (Vienna) and return in time for their trade mission. We drank very expensive cups of coffee for hours until our British Airways plane took off. Such problems were common in socialist countries; it was an advantage to be an official delegate.

Frequent visits to Mexico (sometimes with Jessie) sounds like fun, and sometimes it was: mariachis; balloons in the Alameda; archaeological sites in Yucatan, Tula, and Jalisco; Las Mananitas in Cuernavaca; the Sondheimer villa in Acapulco; presidential receptions; and smiling Mexicans. Often it was not fun. A typical trip before jets were introduced (1958–1961) involved 14 hours to reach New York via Goose Bay and Boston, then Fort Worth–Dallas, San Antonio, and Mexico City. If all went well I arrived after about 30 hours of travel on a Sunday night, with 7 hours of jet lag, at an altitude of about 7000 feet. I began work at 9 a.m. Monday. On some occasions I ran into trouble with Mexican visas. Syntex did not pay me, but paid my expenses to visit. Their support raised the problem of what kind of visa I needed: tourist or visitor (requiring five photographs, five sets of fingerprints, and reams of paper). This decision depended on the particular official involved. On one occasion I was so stressed that I accidentally locked the key inside my car at their Embassy in South Kensington, but managed to read the lock number through the window, go to a Vauxhall dealer, and get another key. I used so much money on taxi fares that I had to go via Cambridge to Manchester to borrow enough money from George Kenner to pay for petrol.

Jessie and I beside our touring car, decked with flowers by the locals, in India around 1970.

On another occasion, expecting to fly from Manchester, I had to stay in London because of Mexican indecision. As a result Jessie, who had just learned to drive, had to go 180 miles back to Manchester with three small children, one with travel-sickness. I cabled Syntex, who exercised the appropriate pressure in Mexico. The next day the Mexican officials in London threw everyone else out: "Yes sir, please sir, what sort of visa do you want, sir?" Passing through immigration and customs in Mexico (and other countries) could be an ordeal, until you learned the local ropes, like the true meaning of "special favor, no open bag"! The favor was for them, but they "no open bag" anyway.

I grew a beard so that I could easily be recognized at foreign airports by people who did not know me. The joy of traveling comes into question after endless boring hours in airplanes and airports; delayed flights; missed connections; arrivals with an internal clock entirely off schedule; and difficulties with languages, currencies, meals, and accommodations.

Jessie and I in front of "erotic" sculptures (censored by the camera) in Halibid, Karnataka, India, around 1970.

Most of the time on my world tours I was working, but there were some tourist opportunities occasionally. My wife accompanied me when she could, for example to Nigeria, the United States, Holland, Germany, Russia, India, Switzerland, and Mexico. We were particularly fortunate in Switzerland, where we enjoyed the friendship of Bruno Vaterlaus and of Otto Isler and the hospitality of his historic Schloss Uttwil on the Bodensee. It has Roman cellars and is mostly an 18th Century "great" house (*Schlossli* or *chateau*).

Industrial Connections The proximity of Manchester to the British chemical industry gave it a unique support base and tradition from about 1870 to about 1940. I developed a number of industrial connections there, based on this tradition. The experience gave me a great respect for what I later classified[77] as strategic research and problem-solving. That approach is distinct from uncommitted research, which most scientists like because it gives more chance to "play around". I acquired an appreciation of the high intellectual qualities: originality, broad outlook, and persistence required for strategic work. Indeed, I would now classify my steroid and metal–ammonia work (Research Set 2) as basic strategic rather than uncommitted research.

Industrial connections assisted a number of practical aspects of my own work, particularly those involving biological activities (Research Sets 2 and 3). It provided early access to instrumentation, especially NMR spectroscopy (in Roche, Basel) when this was only beginning to be available in universities. I also obtained some financial research support. I am grateful to people like Frank Rose (ICI), Leon Golberg (Bengers–Fisons, Holmes Chapel), Otto Isler and Bruno Vaterlaus (Roche, Basel), Carl Djerassi and George Rosenkranz (Syntex, Mexico, Palo Alto), and Ian Morton (Unilever, Colworth, Port Sunlight, Vlaardingen) for the insights and assistance that were provided. I applied these experiences later in teaching and in policy considerations.

Research Set 3

Biosynthesis; Biosynthetic Structure Determination; Alterations of Biosynthetic Pathways

The discipline of biochemistry was evolved to examine the chemical and physical characteristics of molecular change in living matter. It is concerned with metabolism, reproduction, and heredity, and it leads a chemist to some understanding of enzyme catalysis in molecular change. Most of the initial biochemical work was of the "finding-out" type.

Most early biochemists, because of their preoccupation with the common factors of organisms, tended to neglect the formation of those secondary metabolites that mark the diversities rather than the commonalities of living systems. The general study of such substances has contributed greatly to organic theories and mechanisms, and to "real" biochemistry in often unexpected and unsought ways. Organic chemical speculations on biosynthetic routes have not only been fun for chemists, but have revealed new paths not foreseen or discovered by biochemists.

As I have discussed elsewhere,[78] the unraveling of natural-product structures revealed atomic patterns in their molecules that could be related to simpler known biochemical structures such as amino acids, succinate, acetoacetate, phenylalanine, and so forth. Winterstein and Trier about 1910, in a publication now difficult to obtain,[79] first tried to extend the formal paper relationships of structures to actual biosynthetic events. They suggested the correct origins of the benzylisoquinoline alkaloids from substituted phenylethylamines and arylacetaldehydes, both derived from the phenylalanine series. R. Willstätter, from about the same period, explained the origins of, for example, the cocaine alkaloids, in terms of appropriate intermediate molecular reactivities to such structures from known possible simple precursors.[80]

This approach assumed, as stressed particularly by Robinson in 1917,[81] that Nature operates synthetically by rules recognizable to the chemist, despite the very specific influence of enzymes on the asymmetry of products. He arrived at this conclusion following the success of his tropinone (cocaine-type) laboratory synthesis in water at room temperature, from succindialdehyde, methylamine, and acetonedicarboxylic acid. He submitted this paper a week before the theoretical paper pointing out that such intermediates might be chemically derivable from the pool of general metabolites. Synthesis therefore preceded biosynthetic ideas in this case (not the reverse, as usually assumed[82]). The chemical nature of probable biochemical intermediates, according to his view, should be definable in principle as one of a set of alternatives. Much later, chemical interest led to prediction of the exact biosynthetic sequences of molecular conversions that could be tested in biochemical practice, although Robinson never bothered. After 1945, with the availability of isotopic tracers that permitted definition of the fates of individual labeled molecules, some detailed testing became feasible.

Because of my lack of biochemical facilities, I started with only the chemical interests of structure correlation. I graduated to the fundamental biochemical level as the means to do so could be obtained.

Applications to My Sydney Situation

Strategic research ideas necessitated by my approaching "transition" from Cambridge to the Chair in Sydney affected my interests in late 1951. I looked through biogenetic eyes at the literature on natural-product structures, particularly Australian ones, and came across **58**, which was assigned to the wood product campnospermonol. This molecule has an obvious oleic acid side chain. Unlike standard fats, its chain is joined by C–C to the aromatic ring, not by the –O– of an ester group. Knowing the accepted origin of fatty acids from acetate units, I suddenly saw in my mind a string of them continuing through the side chain into the ring. Could that aromatic ring also be acetate-derived?

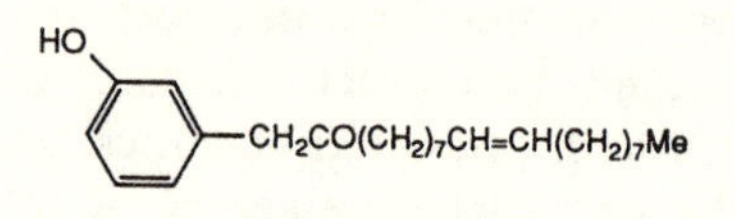

(58)

My approach was partly prompted by my personal awareness of the work of Sir John Cornforth (after he left Oxford) and G. Popjak on the biogenetic origin of fatty acids and steroids from acetate units (acetyl coenzyme A). If such were indeed the origin of this oleic chain, the substitution pattern of the ring might reveal some trace of the original acetate units if appropriate carbonyl groups (finishing as enolic OH) remain as remnant markers of the precursor chain structure. This necessity has to be combined with structural requirements for an aldol ring closure of the chain. The only remnant markers here are the meta phenolic OH and the position of ring closure. Therefore, if my idea of the possible origin of campnospermonol is correct, already further biogenetic changes need to be postulated. These include carbonyl reductions, also related to fatty acid biogenesis, and mechanistically acceptable decarboxylation.

The mere concept of such an "acetate" origin prompted me immediately to draw on paper the simplest β-polyketo acid chain (**59**) that could then be schematically cyclized according to respectable organic mechanisms. There were two obvious mechanistic ways to achieve its ring closure, as I suggested for the first time in 1953.[83] I could apply an aldol closure to the known orsellinic acid (**60**) or a Claisen (C acylation) closure to **61**, the prototype of many known natural acylphloroglucinols.

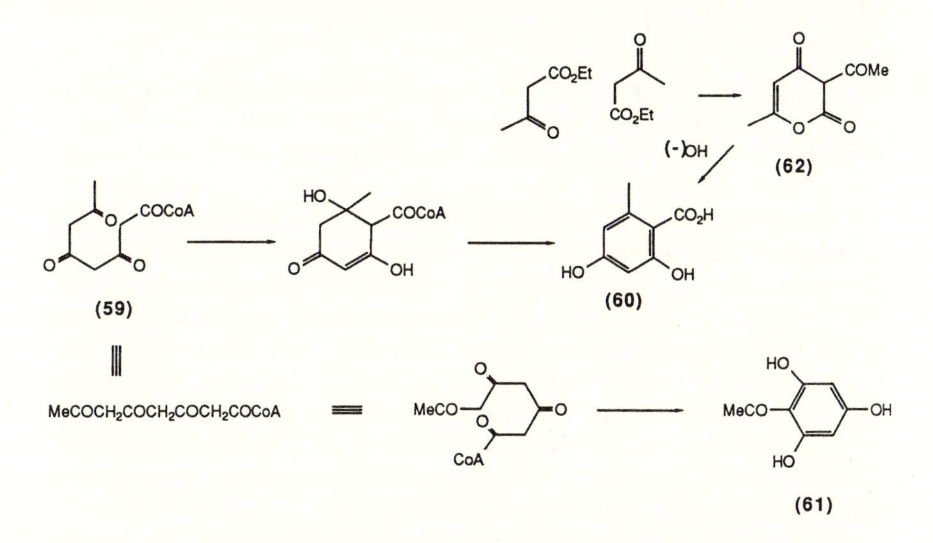

I was delighted and amazed to discover that in a few minutes I could derive on paper many other known natural products by involving more acetate units in longer β-polyketo acid chains, with mechanistically acceptable ring closures of various types. I then suggested for the first time that different starter units for the chains (such as the hydroxycinnamic acids, presumably through their coenzyme A esters) led to the correct complete skeletal origins of many other plant products and to the related plant stilbenes by an alternative ring closure. I discuss elsewhere examples like the flower-pigment flavonoids and anthocyanins through the chalcones.

Within a few minutes these ideas reduced to order what had previously seemed an unrelated jumble of structures of many natural products. It was an immensely satisfying and emotional moment.

The dazzling simplicity and generality of the new correlations can be seen by contrast with two more or less contemporary unsuccessful attempts at correlations by experts: T. A. Geissman on plant products and Harold Raistrick on mold products.[84] Apart from a few generalizations like the C_6C_3 and C_6C_2N (phenylalanine–tyrosine) ideas of Winterstein and Trier (used by Robinson) and the isoprene basis of terpenes (then purely formal, not a direct biogenetic correlation of Semmler and Ruzicka), the structure–biogenetic correlations examined in contemporary reviews contain little of direct use in suggesting structure correlations or crucial biochemical experiments. Neither are they intellectually satisfying, as a jumble of unrelated ideas.

Historical Background

As I was coming into the area without any research experience, my hypothesis seemed to me a totally novel approach. However, during discussions in Cambridge (late 1951), Percy Maitland said, "There is nothing new in that; J. N. Collie suggested the idea in 1907." (He had known Collie, as had Robinson.) I perhaps overreacted initially to give Collie credit, but on further consideration of all the published literature, I question whether that statement is true in any real sense. Collie certainly postulated polyketides, but not acetate units, despite the chemical origin of his hypothesis in dehydroacetic acid. He appeared to favor the direct origin of polyketide chains from carbohydrates. I have discussed the related ideas in more detail.[78] My suggestions, and his, need examination within the general context of how ideas arise, how they are effectively expressed, and how they influence the development of a topic, if at all, which Collie's did not.

Collie had accidentally converted into orsellinic acid (**60**) the synthetic dehydroacetic acid (**62**)[85] (for which he had an incorrect structure, although that is unimportant). By extrapolation of the precursor origin, he suggested in 1907 that **60** might arise naturally from two acetoacetate units[85] (not from one eight-carbon polyketide chain, as I independently suggested in my hypothesis). He published no details of the extension of his ideas to other natural products, although he implied that it could be extended.

So far as I can determine from an extensive authorized discussion in a book by his friend A. W. Stewart,[85] he thought his polyketides must come biochemically directly from sugars, not from acetate derived from acetic acid units. The only other examples mentioned by Stewart, like benzylisoquinoline alkaloids, do not fit naturally into that pattern. They require molecular contortions, which include postulating the availability of any length of sugar chain and any required enzyme that could carry out any reduction, oxidation, or dehydration desired, thus effectively neutralizing any predictions. It is not surprising that nobody took any notice. Apart from orsellinic acid, Collie did not discuss even one genuine polyketide. His discussion (as reported by Stewart and himself) is muddled. Sometimes he seems to think that sugars arise from polyketides, sometimes the reverse, but nothing is definitive.

Robinson, in his only mention of the ideas, published as late as 1948,[86] erroneously correlated the occurrence of the four-carbon polyol erythritol with acetoacetic acid and orsellinic acid in a lichen. He even countered the correct suggestion of Winterstein and Trier[79] on the benzylisoquinoline alkaloids (from C_6C_2 and C_6C_3: phenylalanine-derived) by invoking an unknown type of C_{15} sugar, possibly following Collie.

The Role of Faith

Collie also indulged in some unrelated theoretical eccentricities (e.g., on the structure of pyrones). Although not totally incorrect, they were not acceptable to his contemporaries and probably contributed to the lack of confidence in his general ideas. If he had been right in his biosynthetic postulate, we would pay tribute to his unreasoning faith that most natural products are polyketide-derived from sugars, but in the cases he actually discussed he was wrong. It is sad that the baby was thrown out with the bathwater. If his idea had been properly circumscribed, and if anyone had believed in it, it could have been used since 1907 as a kind of structure-defining (isoprene-type) rule, as I first used it in 1955–1956.

The Collie affair is a classic example of how to guarantee nonacceptance of a brilliant idea by not setting appropriate boundaries and by presenting it in the wrong fashion. Collie was a very interesting man, a celebrated mountain climber; there is a peak in the Rockies named after him. In this activity he resembled Robinson, who took his second wife, Stern, for a honeymoon on the Matterhorn, to her dismay, she told me. They were both scientific mountain-climbers for their own satisfaction, but often that is not enough.

Testing My Hypotheses under Difficulties

In Sydney in 1952 I had not the means (or the techniques) for an experimental attack on biochemistry. Therefore I analyzed structures on paper to support an hypothesis I already knew instinctively to be true. A broad relevant type is exemplified in the formula of one (63) that I contemplated among many lichen depsides (mold metabolites). It led me to a statistical summary of all of the types of substituents found in the aromatic units of recorded structures of depsides and depsidones (64).[87] To avoid personal bias, I took without selection all the formulae from an authoritative review by Nolan.[88]

Apart from the presence of extra C_1 units (to which I revert) and some extra oxygenated substituents clearly introducible by aromatic *o,p*-oxidation processes of known types, the substituents are of the right kind. Typical examples are odd-carbon side chains with a *β*-keto group (if any) and OR substituents that are mostly in the right nuclear positions meta to each other or to ring closures (compare **60** and **61**), with some O missing for reasons to be considered later.

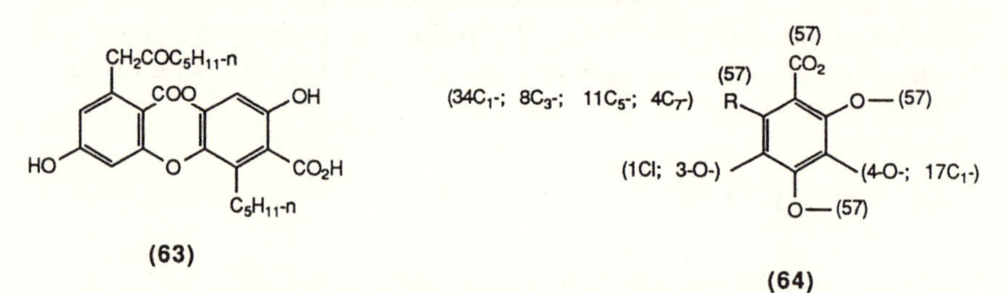

(63)

(64)

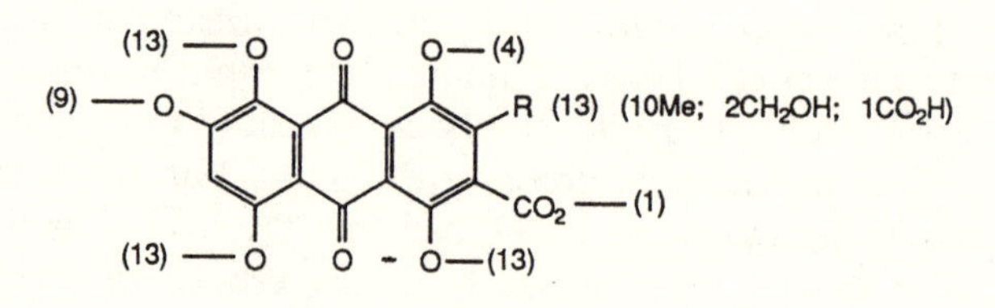

(65) POLYKETIDE

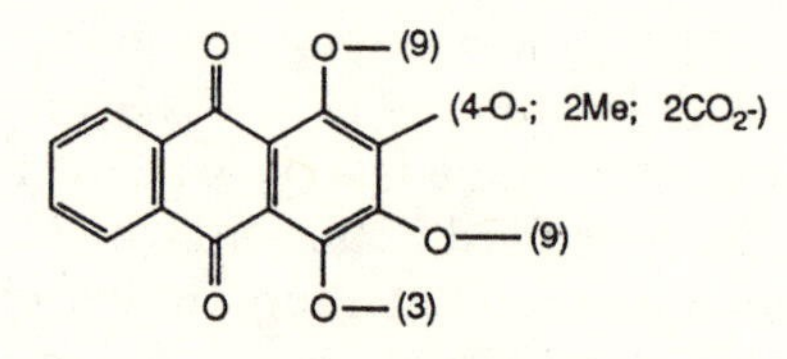

(66) NONPOLYKETIDE

I did the same statistical analysis with natural anthraquinones[89] and showed two substituent distributions, one polyketide (**65**) and the other (**66**) biogenetically different. This distinct origin was later biochemically confirmed by others. Apparently Asahina had previously noted the odd-numbered side chains in depsides, but had made no biochemical correlation. My statistical presentation pinpointed clearly the great extent of the structural coincidences, but also the pressing need to explain several divergences such as extra C_1 units present in depsides and some oxygen-pattern differences in anthraquinones.

From the Complex to the Simple: Protective Hypotheses

Significantly, as with many hypotheses, this hypothesis began with a complex example, campnospermonol (**58**).[83] The biochemical explanation of its structure already required two protective hypotheses on top of my original one: reduction and decarboxylation. Another protective hypothesis, extra C methylation,[78] became necessary on inspection of **64**. I gave lectures in Birmingham and Liverpool on these ideas in early 1952, before leaving for Australia. My friend F. M. Dean, who attended, pointed to the anomalous presence in many lichen products of extra carbon atoms, as I had already recognized. He thought that this invalidated my hypothesis. My reply was "Let us not take too much notice of the evidence. The degree of coincidence is such that I am sure that such anomalies will be explained."

Shortly afterward I proposed C methylation from methionine, to the scorn of the biochemists I approached, as it was then an unknown biochemical process. I was later able to provide the first confirmation of this biochemical reaction with mycophenolic acid (**71**) (O and C methylation from a ^{14}C source yield high and equal labeling, as shown)[90] and griseofulvin (**72**). I followed this demonstration with many other biochemical examples. In this case, as in many others, structure analysis preceded and predicted biochemistry.

Why Was Biochemical C Methylation an Inevitable Hypothesis?

The bases of my novel idea were

- the known biological methyl cation behavior of biochemically active methionine (a quaternary S-Me derivative)
- the structures of the natural phloroglucinols such as **7** (*see* page 65; the β-triketones on which I had worked structurally in Cambridge had methyl alternatively present on C and O to varying degrees)

- the knowledge that laboratory reactions with MeI (as a Me cation donor) can synthetically produce both O and C methylation in many enolic structures, especially in (polyketide) acylphloroglucinols, to yield the types of natural structures I had defined as "β-triketones"

As the result of our subsequent biochemical experiments, natural C methylation from methionine was shown to extend well beyond the acylphenol series and related structures. We defined it experimentally, for example, in simple phenolic (shikimate-derived) rings and in a potentially activated (enolic) carbohydrate unit, in the antibiotic novobiocin (**85**).[91] C methylation of the side chains of steroids and triterpenes from methionine was to me a mechanistically surprising process discovered by biochemists using random finding-out methods. But knowing their result, I suggested in 1960[92] that side-chain C_2H_5 found in situations similar to that of Me in steroids and triterpenes results from a doubling-up of methionine carbons. The origin of this side chain was later discovered biochemically, apparently without knowledge of my suggestion; biochemists do not appear to read chemical papers. We also showed that the tetracycline antibodies arise predictably from acetate and methionine.[93]

Missing Oxygens: What Mechanism?

Another feature in many of the polyketide phenols is the lack of an oxygen substituent in an expected position. This lack is exemplified by our first biochemical vehicle, methylsalicylic acid (**67**), when compared with the fully polyketide orsellinic acid (**60**). As I suggested in 1953,[83] this lack must be due to reduction of a carbonyl into a carbinol group in an intermediate, with dehydration rather than enolization, finally resulting in a stable aromatic ring with an oxygen missing. At which exact stage of a sequence (in ring-open or aldol-ring-closed, but nonaromatic material) this reduction and dehydration occurred was then not evident

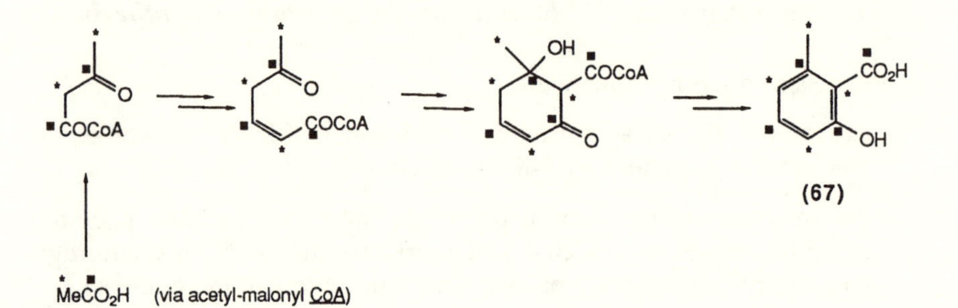

from purely chemical considerations. I inclined, erroneously as a chemist, to dehydration after ring closure, thus avoiding the difficulty that dehydration in a ring-open precursor would need to give specifically a cis- double bond (as later biochemical investigations do indeed demonstrate) to **67** as shown. However, it was clear that sufficient carbonyls must still remain within a chain to activate required ring closures to cyclic, eventually aromatic, final metabolites.

Why Extra Oxygen Substituents?

I readily accepted the occurrence of metabolic introductions of oxygen into reactive enolic positions, in either ring-open or phenolic precursors. It agreed with then-known biochemical processes and chemical reactivities to form phenols or quinones with extra introduced OR or quinone substituents, such as in flaviolin (**68**). I foresaw that this process could be experimentally confirmed by using oxygen isotopes, but we did not have the practical means to do so.

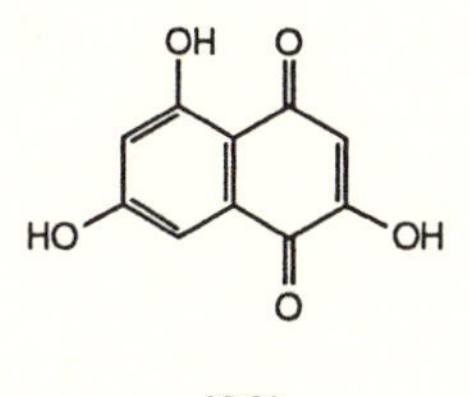

(68)

Propionate as Well as Acetate?

I discussed my polyketide hypotheses internationally at a Gordon Conference in 1954. Contrary to Robinson's statement in his 1955 book[94] that "the hypothesis had been in the minds of organic chemists since Collie" (1907), none of the distinguished audience had ever heard of it. What Robinson really meant was that he knew about it; nevertheless, he did nothing with it.

I had a constructive discussion at the conference with Bob Woodward, who was then working on the partly determined structure of the nonaromatic macrolide antibiotic Magnamycin. In my view, from what was known, it was undoubtedly of polyketide origin, but it contained a number of extra Me branches. I postulated the origin of such branches in methionine plus acetate. Woodward then proposed propionic acid as a chain-building unit (mainly, I think, because of terminal Et present). He turned out to be right in this series, although I was right in just about every other. Both routes can indeed be involved in some individual compounds: Ockham's razor can be a blunt instrument in biology. When supporting proof was later obtained in this series alone, I said, "Bob, at last we have experimental proof that you have made the Faustian bargain!"

Testing

In 1952 in Sydney, with no money and no equipment but with some enthusiastic students, what evidence could I gather to test my hypothesis? A beginning on the task lay in successful biogenetic structure predictions of eleutherinol (**69**) and flaviolin[95] (**68**).

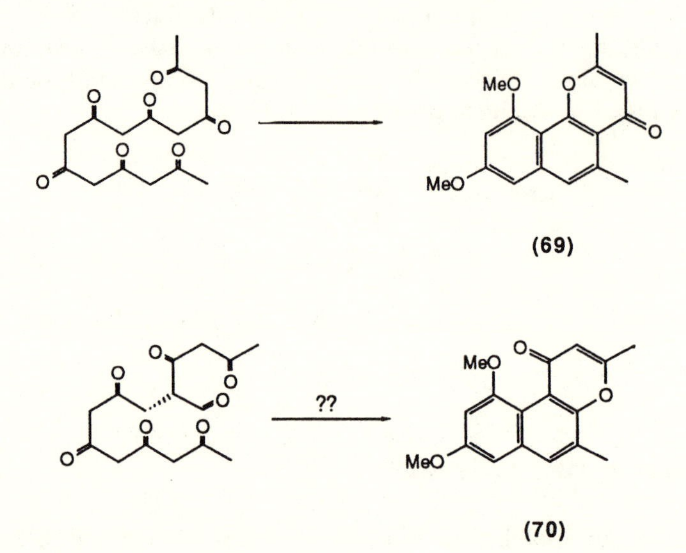

One case involved correction of an assigned chemical literature structure **70**, which I questioned on the basis of my postulated origin from one polyketide chain (from its partial, known, substitution patterns). Such origin could not contain the necessary ring-closure activation throughout, as shown. I correctly assigned the structure on that notional basis (**69**).

My second case (**68**) represented a final correct definition of a structure not fully demonstrated by chemical evidence. I placed an unassigned OH (in the quinone ring) on the biogenetic grounds of a postulated polyketide origin followed by a quinone oxidation. The structural evidence was all that was then known of its oxygenated positions and ring-closure, which I interpreted correctly. Several people then unkindly pointed out to me that in each case I had a 50% chance that my conclusions would be correct. However, many other examples followed and no subsequent exceptions were found to my general reasoning based on statistically acceptable patterns. Some evidence must exist that a particular biogenetic rule applies.

These novel uses of partly known structures are reminiscent of the older structural isoprene rule for terpenes, to permit the valid choice of a more likely final structure for which chemical evidence alone is ambiguous. In such cases reasoning rests in chemically determined substitution patterns. Later I was to develop, in more difficult cases, direct evidence for biogenetic origins by incorporations of isotopically labeled precursors.

Terpenoid Cations: What Basis?

I was also forced to suggest, on the basis of structural analyses comparative with C methylations, similar biochemical introductions of terpene C_{5n} cations on to carbon. This suggestion was made before the discovery of mevalonic acid and its phosphates. I then contemplated and synthesized terpenoid analogs of methionine[96] as biochemical sources of terpenoid cations. This process turned out to be one of my few incorrect guesses, but we did not waste much energy on it.

Radiotracer Testing

The first tracer support for my basic polyketide hypothesis, which then became a theory, was obtained in 1955 in Sydney. The biochemical origin of 6-methylsalicylic acid (**67**) from four 1- or 2-^{14}C-labeled acetate units[97] was confirmed by using degradation methods we devised. The detailed sequence shown is now known with some certainty; we could not then examine at which stage reduction occurs. Later, in Manchester, we confirmed the skeletal origins of many other structurally diverse fungal compounds involving different numbers of acetate units and different structural types, such as the important antifungal agent griseofulvin (**72**). Sometimes, but rarely, postulated origins required involvement of more than one polyketide chain.[98]

We predicted, and confirmed, secondary reactions like ring contractions (e.g., in the polyketide cyclopentanoid terrein, specific ring openings of aromatic polyketides (e.g., the heterocycle penicillic acid),[99] and the involvement of other polyketide chain-starter units and introduced units, particularly of terpenes and degraded terpenes.

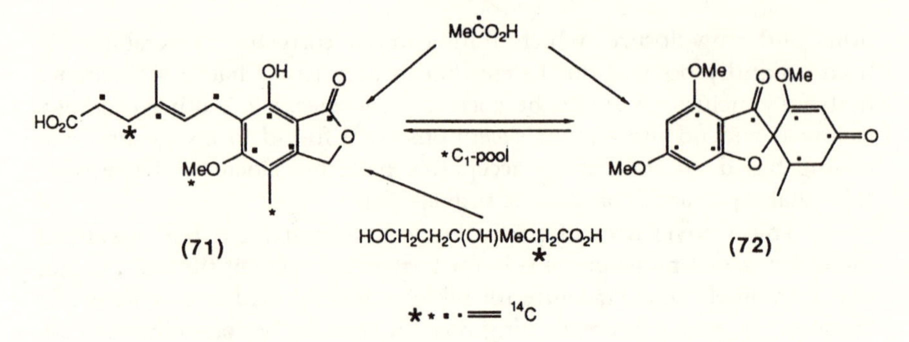

One outline example I give of mixed origins (acetate–terpenoid–methionine) is of the compound that we spent more time on than any other, mycophenolic acid (71).[100] Later workers have detailed the exact size of the initial polyterpene side chain in its precursors. This substance was probably the first pure mold antibiotic ever isolated (identified by Gozo in the 1890s). Obtaining it in quantity is now of great pharmacological interest because of its immunosuppressive activity.

Why Bother; Why Not Just Find Out? The fun of such work was to pick structures, often already partly known, analyze their possible origins on paper, and confirm the correctness of our analyses in the laboratory by seeking experimentally to show we were wrong, although we never were. Finding out biochemically is less fascinating intellectually. In some unpredictable cases we were led, successfully, to discover novel biosynthetic routes (but even that success was guided by instinct).

What Are Acetate Units? At the biochemical level, I initially recorded my view in 1953[84] that acetate is acetyl coenzyme A. Later experimental anomalies about the exact extent of incorporation of the labeled first acetate unit[98] into a chain, observed by my collaborator Rickards, suggested that the starter acetate unit of a polyketide was biochemically distinct from the others. Later, other workers showed by biochemical work that for incorporation as a chain-building polyketide unit (not a starter unit, which, as we suggested, could in principle be any natural acyl Co-A), acetate is reversibly converted into malonyl coenzyme A and used as such. Thus Nature is a good organic chemist, using the reactivities of the S-coenzyme as we initially suggested,[84] but also using the reactivities of the introduced extra carboxyl to generate a necessary carbanion. The chain-labeling difference we observed provides a method of distinguishing the presence of more than one polyketide chain[101] in some molecules.

Extent of My Hypothesis. At least 4000–5000 natural products are now known or believed to be fully or in part polyketide in origin.

Other Molecular Types and Origins. Some important natural compounds, among many whose origins I first predicted and then defined biochemically, are tetracycline antibiotics (acetate plus methionine N-Me and C-Me),[93] the aromatic basis of the ergot alkaloids, such as structure **79** (mevalonic acid and tryptophan)[102] and the flavonoids and anthocyanins (cinnamic acids and acetate) (to be discussed in more detail).

Biochemical Progress. All of my earlier biochemical experiments (1956–1968) involved the use of radioactive ^{14}C, with degradation-activity measurements to pinpoint molecular labels. My collaborators did not like this boring and exacting work, although they liked the results. With the development of NMR and the availability of isotopes like ^{13}C and ^{15}N, it became easy to examine labeling patterns spectrally. This newer approach, which we did not pioneer but later used, has a further great advantage: the exact orientation of the CCO in polyketide chains and other units can be determined. This information cannot be obtained by the radiotracer technique.

What Is in a Name: Why Polyketide?

Chemists are often rather fanciful and even inaccurate in conferring names. The term polyketide, suggested by Collie, is fully indicative and limiting on the class (origins in chains with CO in alternate positions, no matter how generated). For some unknown reason the term acetogenin was later coined in the United States without explanation or apology. It is wrong on three grounds: language, because the series does not generate acetate and the ending "genin" has a well-defined and quite different chemical significance; chemical, because the defined class then irrationally excludes not only substances like fatty acids and terpenes (derived from acetate, if that is the criterion), but also some fully analogous polyketides because they happen to arise from propionate or other units; and of course there are grounds of historical precedence of naming. I realize that people who indulge in naming can often claim originality of a concept.

I do not quarrel with names, unless they are misleading or irrational. But why some have survived and others have been altered is puzzling. The original and correct "anionoid" and "cationoid" classification of reagents by Lapworth and Robinson, for instance, has been displaced by the later and less correct "nucleophilic" and "electrophilic" of Ingold (all bond formations involve electrons). Taste or energy can sometimes prevail over sense. I recall Clarence Smith (editor, Chemical Society) quarreling with Syd Plant about an arsenic series named by analogy with the azoles! "I am not having words like that in the *J. Chem. Soc.*" Some natural-product chemists name new compounds after spouses or significant others, which (like tattoos) may later prove embarrassing. Barbiturates were actually named, as Willstätter[80] records, by von Baeyer after his friend Barbara. (Did she send him to sleep?)

Terpenoid Biosynthesis

In Sydney in 1952 I was looking in eucalyptus leaves for the then-elusive unknown precursor of their terpenes, which I suspected to be an acid. We obtained instead shikimic acid[103] in great quantities as the second isolation from a plant (eucalypts construct lignin very rapidly). Our mold work, rather than our limited higher plant work, gave many opportunities to examine the biogenetic aspects of terpenoids. I have noted some terpenoids of mixed biogenetic origin, including some like mycophenolic acid (**70**) with altered or degraded terpene chains.

Some molds were known to produce cyclic mold–plant diterpenes, such as rosenonolactone (**73**), and also the important plant hormone series gibberellins, such as structure **74**. Although the gibberellins are a C_{19} series, I was the first to suggest a degraded C_{20}-diterpene by interpreting what was known of the partial structures contained in them. These substances were the first biochemical examples examined by tracer methods for a diterpene origin.

We confirmed our expectations and elucidated many details of chemical changes predictable on theoretical grounds, of ring closures, contractions, and stereospecific Me migration, consequentially on mechanistically predictable concerted cationic cyclizations and rearrangements of a specifically folded diterpene chain. Some postulated and later confirmed labeling is shown from **75** to **73** and **74**. To accomplish this, we incorporated 2-(^{14}C)mevalonate and 1- and 2-(^{14}C)acetic acid and examined the positions of labels in the products.[104]

One of our major experimental tools was the Kuhn–Roth oxidation with chromic acid, forming acetic acid from molecular C-Me. The technique was developed by Cornforth and Popjak in their previous steroid work, and they communicated it personally. This method permitted us to examine, for example, the radioactivity of the totality of Me groups and of the carbons to which they are attached. In a molecule derived from precursors like 1-^{14}C-acetate, the results were particularly

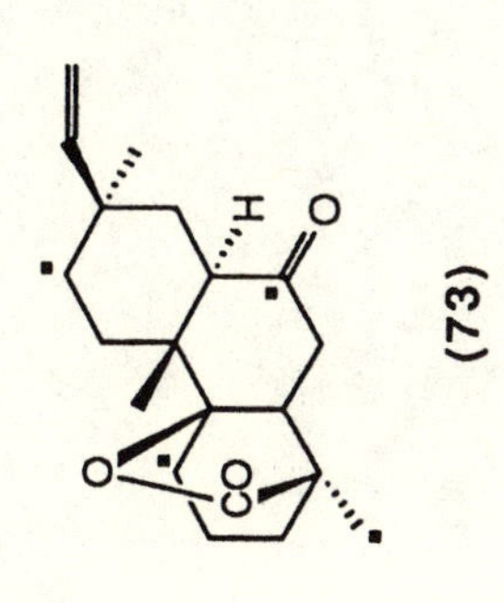
H
O
CO
O
(73)

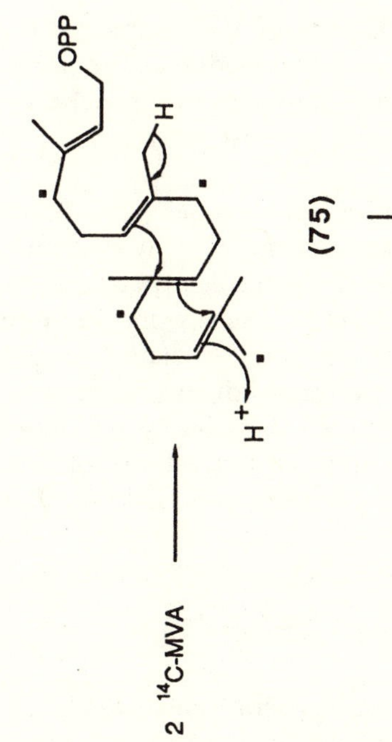
2 ¹⁴C-MVA
OPP
H
H+
(75)

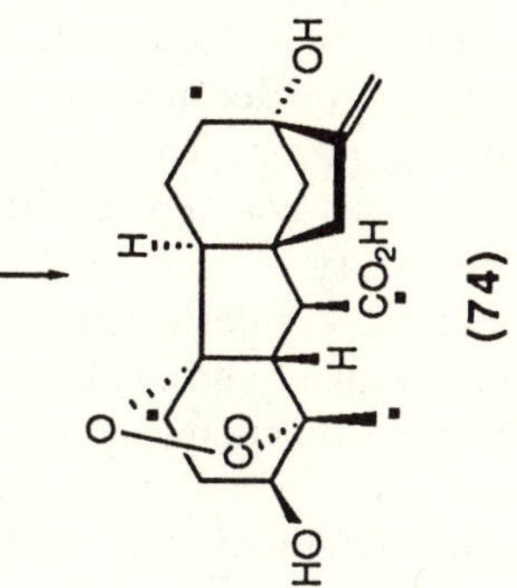
OH
H
CO2H
H
O
CO
HO
(74)

Sir John Cornforth (seated) and I at an IUPAC Natural Products Symposium in 1960.

important. They gave the totality of labeling of Me groups and the carbons to which these are attached. From a knowledge of the number of Me groups, many deductions could be made about biogenetic cyclizations and migrations. This method is now primitive in relation to NMR techniques, but it was critical then.

The theory of concerted cyclizations of terpene chains was advocated by Stork, Wenkert, and others at this time[105]. It was attractive to the organic chemist and in line with what should be expected from sterically specific biochemical ring closures. This theory envisages a "zipper" effect of a cation through a folded terpene chain (e.g., **75**) producing a fully defined stereochemistry in the products (differing according to the type of chain folding, as defined particularly by Duilio Arigoni and Alfred Eschenmoser). We confirmed the validity of this general theory through ready tracer incorporation into *Gibberella fujikuroi*, obtained from ICI Pharmaceuticals.

Labeling in Terminal Me Situations of Terpene Chains

But what we needed for certainty in our general proof was to confirm another expected but unproven major biogenetic feature in open-chain terpenoid precursors. Our expectation was based on the general grounds of enzymic stereospecificity and retention of configuration. In the terminal chain *gem*-Me_2 attached to a double bond, only the Me that is trans- to the main terpene chain should become labeled by 2-^{14}C-mevalonate precursor, because it should arise selectively from that 2-position.

We proved this expected pattern laboriously, in the absence of the present relevant physical methods, by examining the open-chain monoterpene hydrocarbon **76** that was obtained by our metal–ammonia cleavage of the mold product mycelianamide (**77**), a reaction we used to define its structure.[100] After feeding the mold with 2-(^{14}C)mevalonate or acetate, we determined the expected labeling pattern in this hydrocarbon as shown. (* is ^{14}C from the 2-position of mevalonate; the other labels are from the 1-position of acetate; we also confirmed the converse pattern with 2-labeled acetate.) Standard Kuhn–Roth degradation to acetic acid demonstrated that it contained, as expected, exactly two labels (*). One label is on a carbon of $=CMe_2$, but on which of the two sterically different methyls?[106]

Chemical Cheating: Why Use Rabbits?

This labeling question was settled by feeding the labeled reduction hydrocarbon to rabbits and collecting the Hildebrandt's acid (**78**) from their urine. NMR easily demonstrated the stereochemistry of this dicarboxylic acid.[106] This procedure is chemical "cheating", according to the rules of the game, but then it was then our only possible approach. We had used enzymic specificity to distinguish sterically different Me.

We had worked on what happens to terpenes in leaves eaten by Australian sheep, so I knew something of the metabolism of terpenes. Rabbits had been used to obtain Hildebrandt's acid in studying metabolism of the corresponding terpene alcohol geraniol, so this approach seemed worth a gamble. We were lucky; not only was the hydrocarbon metabolized, but the oxidation was totally specific for the Me in which we were interested. Later there was other indirect evidence. For example, we determined a similar nonrandom pattern while confirming our predicted origin[102] for lysergic acid (**79**).

Chemical–Biochemical Interactions

A concerted cationic cyclization of a terpene chain would result in nonrandomized *gem*-Me_2 carbons in the first cyclic product. If biogenetically true, we can predict from the labeling we had shown in the open-chain terpene that the equatorial Me in a cyclic diterpene should be the only one of the two to be labeled by 2-^{14}C-mevalonate. The problem with investigating any hydrocarbon series is that the Me could not then

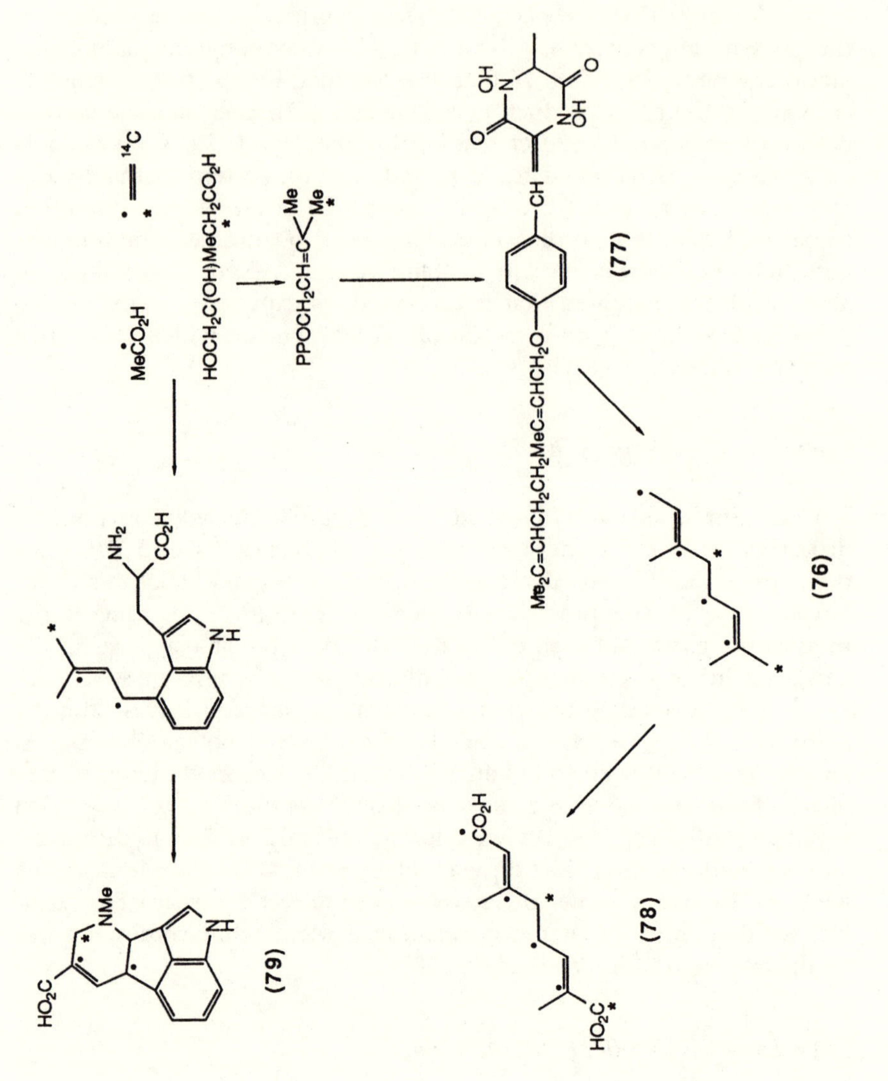

MeCO2H
14C
HOCH2C(OH)MeCH2CO2H
PPOCH2CH=C
Me
Me
Me2C=CHCH2CH2MeC=CHCH2O
CH
OH
N
O
N
OH
O
NH2
CO2H
NMe
HO2C
(77)
(76)
(78)
(79)

be experimentally distinguished (they can now by NMR techniques). We used the fact that one of these sterically different methyls (the axial one) has been chemically differentiated by biochemical oxidation to carboxyl rosenonolactone (**73**) and in gibberellic acid (**74**). The expected labeling could thus be investigated, and we confirmed it.[104]

This result led for the first time to the important generalization that such polyterpenoid cyclizations are concerted ones. Many deductions concerning the origins of specific structures in the cyclic polyterpene series could then be made with full confidence. These include configurations of ring junctions, the steric course of bond migrations of the Wagner–Meerwein type, and of suprafacial Me such as we found in rosenonolactone (**73**). Our many labeling experiments confirmed all of our theoretical expectations. Previously much of this theory (by Arigoni, Stork, Wenkert, and others) was a likely speculation, but it was henceforth on a firm basis. I refer elsewhere to one application with pleuromutilin.

We showed also for the first time that the C_{19} gibberellic acid (a major plant hormone) is actually a C_{20}diterpene that has lost a carbon.

Some Other Hypotheses

Totally different sets of biogenetic areas, which I cannot discuss in detail here, were in part brought to my notice through invitations to participate in CIBA chemical–medical symposia. One novel idea,[107] concerned with the N-oxidation of peptides, led to our experimental verification of the direct origins of acylhydroxylamines from peptides and to suggestions concerning the origins of azaserine, the antibiotic 6-diazo-5-oxo-L-norleucine (DON), and penicillin. My suggestion at that conference in 1958 of the origin of the penicillin nucleus, aminopenicillanic acid, led John Sheehan to a published discussion statement that later, I am told, won him a major very profitable industrial patent lawsuit.

Others of my many hypotheses based on chemical reactivity considerations are worth mentioning. The same type of structure–reactivity analysis as led to the polyketide ideas generated the first suggestion of the origin of the dimethylbenzene ring of riboflavin from two moles of an equivalent of diacetyl. We followed up this study by a biomimetic synthesis of lumiflavin[108] from an aldol dimer of diacetyl. This work was later extended by H. C. S. Wood to a synthesis of riboflavin accomplished by using the correct initial side chain. The actual pathway of biogenesis seems rather more complex than we thought, but the suggestion is fundamentally correct.

An example of another kind of factual correlation of biochemistry with rational organic mechanisms concerns the origins of the natural allyl- and propenylbenzenes, such as safrole, eugenol, and anethole. These important flavor constituents are related to the substituted cinnamyl alcohols, which are lignin and lignan precursors. The C_6C_3 nucleus immediately suggests members of the biochemical substituted phenylalanine series as precursors, but how is the side-chain oxygen lost?

The solution was suggested to me by the existence of various physiological forms of *Cinnamomum* species containing cinnamic aldehyde (or cinnamyl alcohol or acid, with oxygen on the side chain but none on the benzene ring) or safrole or eugenol and congeners (allylbenzenes with no side-chain oxygen but all with at least one aromatic 4-oxygen substituent). The biochemical reductive removal of a side-chain oxygen is apparently made possible by the presence of a 4-oxygen in the ring. This fact mechanistically suggests initial reactive ionization into a stabilized side-chain allylic carbenium ion and reduction by a hydride equivalent (as provided by known coenzymes) into either an allyl- or propenylbenzene. (Both types occur in Nature.)

We provided model support by the formation of mixed safrole and isosafrole by the reduction of methylenedioxycinnamyl alcohol (**80**) with aluminum chloride and lithium aluminum hydride. It was the first use of that reagent consciously based on the intervention of a carbenium ion followed by hydride addition.[109] This sequential origin of structures was later confirmed biochemically in plants by others.

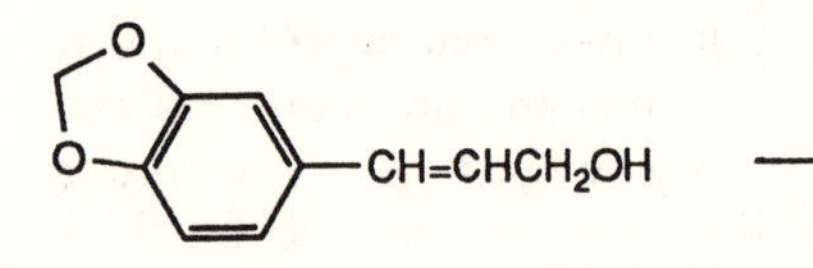

(80)

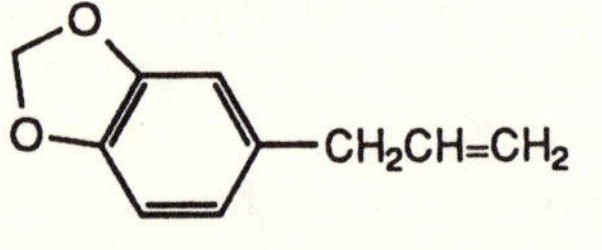

+

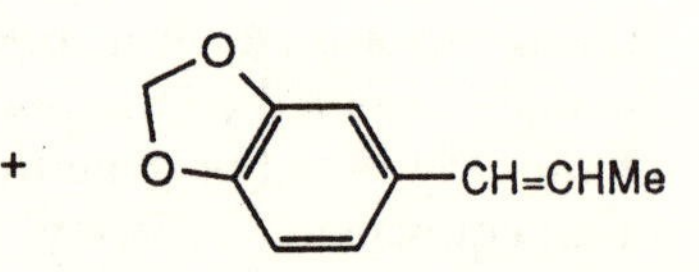

A further hypothesis, based on structure analysis of an extended series of aromatic compounds combined with my ideas of mechanism, is the generation of many natural furanobenzenes from initial introduced terpene units (based on structure analyses), together with a rational mechanism for the loss of the 3-C portion,[110] for which we provided laboratory models. Such an origin was later confirmed biochemically by others.

Wearing Biogenetic Spectacles

I arrived at all of this, and other results not mentioned, in the first instance simply by wearing biogenetic–organic–mechanistic spectacles when looking at structures of any reported natural product or its congeners, and at general structural relations of natural products. This approach was new in 1955 outside the alkaloid area, but slowly became popular after our first publications. Before that time papers on natural-product structures, except alkaloids, usually did not mention any possible biogenesis. Without doubt, authentic results and ideas would ultimately have been attained by "finding-out" methods. Nevertheless, our "artistic" approach is emotionally and intellectually satisfying to those involved and to observers. In addition, it is a demonstration that a superior way to reach the truth is first to dream up the critical experiment. In this biochemical connection, it was difficult to induce laboratory workers in limited classical fields to take any notice of speculations, no matter how firmly they were supported.

My own reaction to the beauty and genius of the revelations of the oxidative–phenol–biogenesis theory of Barton and Cohen was ecstatic and emotional. The theory was based on organic mechanisms that went a long step, for certain structures, beyond my polyketide ideas, which had defined the origins of many of their precursors. Why did I not think of it? I should have, especially as I was dealing with some of its products like depsides and griseofulvin. It is so *obvious* once seen! But then I was totally obsessed with defining the origins of the fundamental skeletons of natural phenols, rather than with what happened to such structures afterward. Obsessions can be obstructive as well as fruitful.

Biogenetic Structure Determinations

The reverse of the analysis of biogenesis from structure is the use of biosynthetic theories to suggest the more probable of a number of possible structures arrived at by incomplete classical degradation techniques or, by the 1950s and 1960s, primitive physical methods. The technique had been used for many years within restricted classes of compound, most consciously by Robinson, for example, who found it crucial in formulating morphine. Less consciously, it was used first in the form of a structure correlation as the isoprene rule for terpenoids by Wallach and Ruzicka, and was later developed by Arigoni and Eschenmoser as the biogenetic isoprene rule. It was a fascinating progressive guessing game, with a partly rational basis.

Use of this approach requires some structural information to define the biogenetic type, even if this information is merely the C_{5n} skeleton suggestive of terpenoid origin. It usually occurs within a limited range of C:H ratios.

I have pointed out for eleutherinol (**69**) and flaviolin (**68**) how overall polyketide structure can be deduced on the basis of chemical evidence. With the advent of tracer techniques, it seemed to me that much more substantial direct evidence could be gained by radioactive incorporations into mold products (in particular) and examinations of tracer patterns. This technique should indicate what units (terpene, indole, acetate, etc.) are involved. Also, labeling patterns show how the skeletons are constructed by joining such units. Our pioneering work involved molecular degradations using ^{14}C; later, with the advent of isotopes like ^{13}C and ^{15}N in conjunction with NMR, it became possible to obtain proofs much less laboriously by examining spectra.

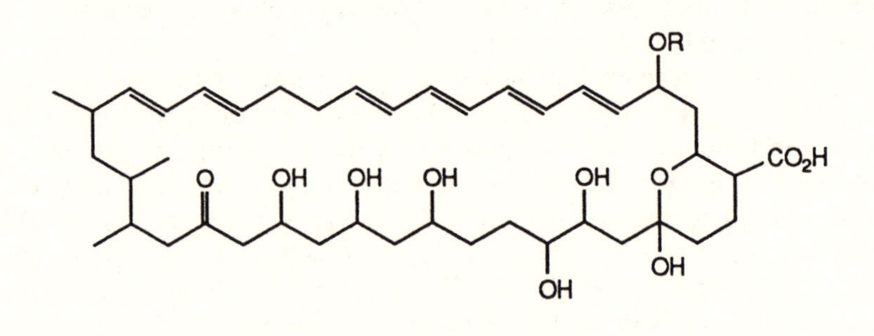

(81) R = noviose

The first example of our new approach was nystatin, an important antifungal agent. My collaborator in the work, R. W. Rickards, finally showed it to have the full structure **81**. At the beginning, major structural features were known largely through the degradation work of Djerassi, our collaborator. From the pieces then known, I thought the structure might be explicable by a polyketide origin of the main skeleton from acetate and propionate, but even the molecular weight was not certain. We fed the producing organism in separate experiments with (^{14}C)acetate and (^{14}C)propionate labeled in all possible positions. The labeling of degradation products was examined, including its distribution (e.g., on a carboxyl or Me of Kuhn–Roth acetate, or in larger products when structurally known). We were thus able to show that only acetate and propionate are involved in the main carbon nucleus, with the numbers of each unit present.[111]

Assuming the mechanistically rational head-to-tail biogenetic linkage of the polyketide units, it was then possible to suggest from defined tracer distributions how the degradation fragments must be joined to each other and therefore how the skeletal pieces they represent combine, on paper, to form the overall molecular structure. Hence, together with minimal other physical and chemical evidence, we suggested the major carbon skeleton to be that shown in **81**. This assignment has been confirmed through further extensive structure work by others.

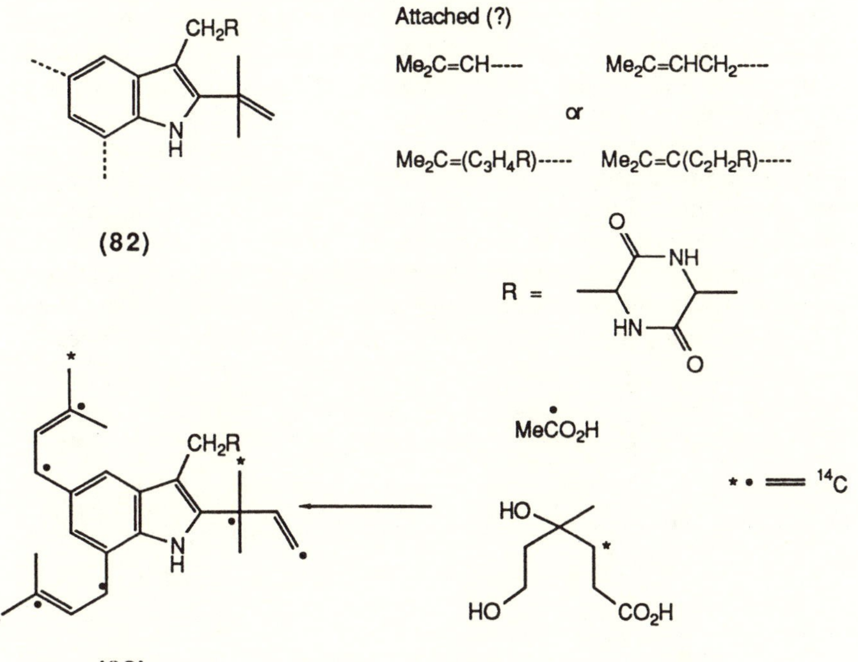

To generalize our basic approach a very different kind of molecule was examined, the mold metabolite echinulin (**83**). When we began[112] its chemically derived structure consisted of the incoherent pieces shown in structure **82**. It suggested to us, although not to the initial workers, a biogenetic origin from a diketopiperazine (tryptophan and alanine) with three introduced C-5 terpene units. To fit this idea, through credible biochemical mechanisms and the chemical evidence, into an acceptable structure **83**[113] required revision of the molecular formula. We added CH_2 and placed three C-5 units in such a way that they could have been introduced by an acceptable carbenium ion mechanism (one as a reversed C-5 unit). Our feeding of labeled precursors to the mold and quantitative examination of degradation products confirmed this suggestion and the new empirical formula. It also showed that the unsubstituted tryptophenylalanine diketopiperazine structure itself is, predictably, a biogenetic precursor.

Novel Types of Structure: Brevianamides

This incorporation work was the suggestive precursor of work on the structure of the brevianamides, the first of a major series of novel fungal diketopiperazines with important toxic properties. We discovered the series by chance in the course of work on the polyketide–terpenoid mycophenolic acid in the organism, *Penicillium brevicompactum.* They formed the first examples of a new important structural and toxic class, with some previously unknown biosynthetic structure generations leading to bridged rings. At the time there were no helpful structural analogies. Many variants on the theme have been discovered since then. Some excellent work has been done in South Africa, notably by my former Manchester collaborator on fungal products, Cedric Holzapfel and his former collaborator, P. S. Steyn, and in the United States.

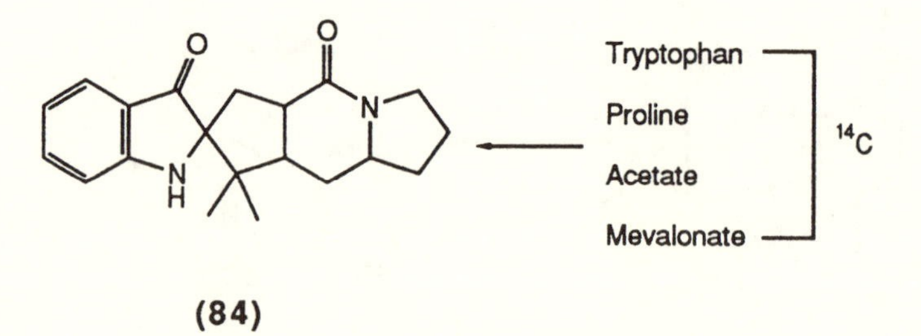

(84)

Our first example, brevianamide-A (**84**), is the most complex of our new brevicompactum alkaloid class, but happened to be the first to be examined because of its availability. Although we had NMR and MS, the condensed ring system and the even distribution throughout it of three carbonyl and three N rendered difficult the identification of overall structures. Parts were visible: CMe_2, a strange type of diketopiperazine probably derived from tryptophan, and an amino acid that was likely to be proline because of the number of rings present from the empirical formula and the lack of Me groups. Accordingly, labeled tryptophan, proline, and terpene precursors were fed and found to be incorporated. On this biogenetic basis it was possible to derive **84**,[113] which readily explained, in retrospect, the MS and NMR spectra. The other brevianamides,[114] or their alternative conversion products, turned out to be biogenetic precursors of **84**. Whether we were lucky or unlucky to have started with the "wrong" complicated congener depended on whether the structures or the methods we had to use were considered more important. These tenuously defined structures were later confirmed by X-ray crystallography.

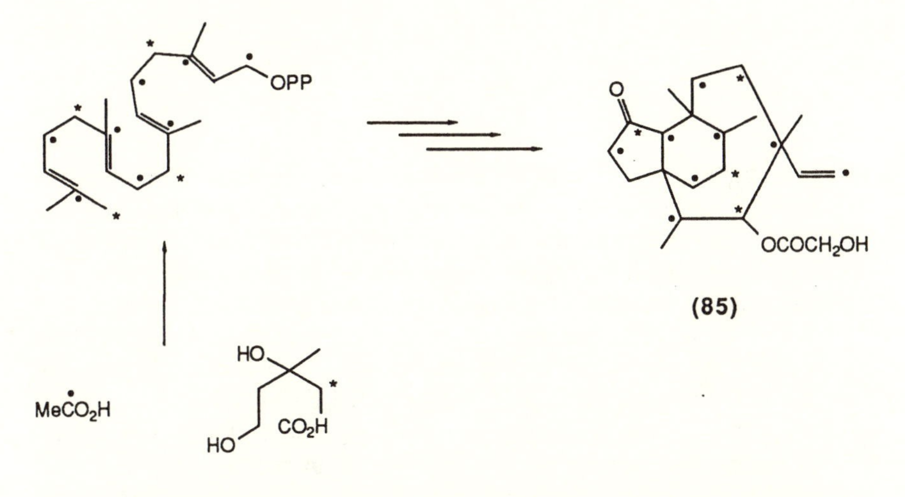

A different type of biogenetic structure method I have already mentioned, of the novel diterpene pleuromutilin (**85**).[115] I further note here that our desire to examine that structure by tracer methods in conjunction with theoretical terpenoid cyclization–migration ideas provided major confirmations, not only of concerted cyclization processes and the suprafacial migration of Me in an intermediate cation within the general diterpene series, but also revealed a new type of further (but explicable) cyclization–migration in forming the final skeleton. In such ways applications of biochemical techniques can also provide further theoretical insights into biosynthetic processes, in contrast to classical mere structure determinations of the compound.

Phomazarin (**86**)[116] was my last structure determined by this approach, using new general technical procedures developed by others, involving NMR with stable isotopes without needing degradations. Its early chemical history caused us to believe that it is probably a complete polyketide. We were able to confirm this readily by using radioactive (^{14}C)acetate and to indicate some crucial structural aspects despite misleading degradation evidence provided by the first workers, Kögl

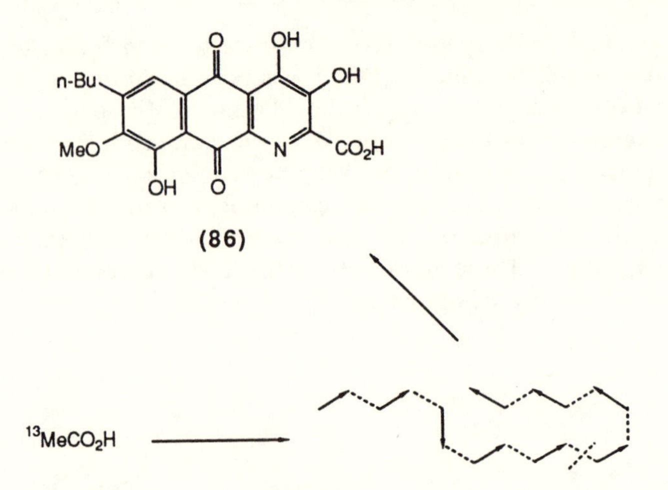

and his collaborators.[117] Kögl's errors were due to premature attempts to use micromethods before the techniques existed, combined with the "evidence" provided by his Nazi collaborator Hänni Erxleben, which he uncritically accepted because it supported what he wanted. This is a genuine tragedy, based on betrayal of trust and gullibility.

The final structure definition was achieved through the first biochemical incorporation for this purpose of ^{15}N (from nitrate), the standard ^{13}C (from acetate), and ^{2}H, followed by examination of NMR couplings.[116] This procedure permitted assignment of the exact situation of the original acetate units within the chain as shown.

But why carry out such procedures nowadays, when rather mechanical ones like X-rays and modern NMR and MS techniques are so powerful and convenient? One answer is that aspects of both biosynthetic pathways and structures are simultaneously illuminated. Also, it provides good broad training in chemical and biochemical research for students. A major reason is that the problem-solving process is much more fun for the researcher.

What Are Biosynthetic Intermediates and What Can Be Accomplished by Feeding Them to Organisms?

The precursor incorporation work, and our resulting knowledge of pathways, led my thinking to radiate from a number of creation centers, usually for very practical reasons. What are biosynthetic intermediates? Are they definable as substances that can be fed to intact organisms, or are they only generated inside cells, or only in adhering to enzymes that pass them on? Can the real intermediates (whatever they are) be fed from outside the cell? Can the production of normal metabolites be increased by feeding such definable precursors, or can new metabolites (such as altered antibiotics) be made by feeding artificial precursors? How far will enzymic specificity stretch in handling such foreign compounds structurally related to normal metabolites? Can mutations that stop a pathway leading to normal metabolites leave intact enzymes that can accept altered added intermediates, possibly catalyzing their conversions into new products? What happens to the products of genetically interrupted pathways in plants; do they accumulate or are they diverted into new pathways? Are chance mutations therefore partial explanations of the occurrences of so many natural products, which are apparently useless or marginally useful to the organisms? Is it possible to pin down in changed biosynthetic pathways the chemical effects of altered genes as reflected by the structures of final products? These are the kinds of questions that occurred to me as a result of the experimental work. I can discuss here only a few aspects of such novel creation points.

First, we were lucky to secure the incorporation of labeled acetate into polyketides. As I realized from the beginning,[83] acetyl coenzyme A, a genuine intermediate to polyketides, does not normally arise biochemically from acetate itself, but rather from pyruvate, succinate, and similar sources. Nevertheless, we were able to feed this simple precursor successfully into polyketides in intact organisms. We were also lucky in the small degree of scrambling of labels in products from such precursors as a result of their general involvement in metabolic cycles. Genuine endogenous biochemical intermediates, such as acetyl coenzyme A or isopentenyl pyrophosphate, often can only be fed into extracted enzyme systems in disrupted cells because such polar molecules may not penetrate cell walls.

I use the term "intermediate" here to describe both compounds that can be fed successfully into intact organisms and authentic biochemical molecules, according to the context.[118]

But are there, in routes to polyketide biosynthesis, any true chemically definable intermediates in the sense of free molecules that can emerge from and be taken up by enzyme systems? So far as I know, no open-chain molecule on the way to a primitive full polyketide skeleton has ever been observed. Nevertheless, secondary modifications of such metabolites once formed is common, and alternative cyclizations of common open-chain polyketide precursors can often be postulated on purely structural and reactivity grounds.

Altered Antibiotics by Partial Biosynthesis: Mutasynthesis

At a conference in 1962[119] I made a new suggestion that was calculated to attempt to modify structurally just-failed antibiotics. Novobiocin (**87**, R = Me) was particularly in my mind then because of industrial connections. I said, "The factors which apply to the stimulation of formation of a natural antibiotic by providing a normal intermediate should also apply to the production of a new antibiotic by feeding a modified intermediate, with an important addition. Competition will occur between the normal endogenous intermediate to give the normal antibiotic and the abnormal exogenous intermediate to give a new one. If the enzymes can deal at all with the new series they might be expected to do so more slowly.... To remove this competition it should in principle be possible to mutate the organism so that it cannot complete the synthesis of the normal antibiotic unless given the normal precursor, which the mutant can use but not make. Provided it can use at all the altered intermediate, the mutant should then be able to complete the formation of the new antibiotic." This principle, since named *mutasynthesis*, has been used in a number of structural series of antibiotics.

Our own successful attempt to implement the idea involved novobiocin (**87**, R = Me). The work was described at the Second International Conference of Biochemistry in New York in 1964, but is still largely unpublished[120] in detail. It was first necessary to define the normal biosynthetic routes, presumably from tyrosine, that we used in 1962–1964; some of the results are schematically summarized.[92] Interesting general points were the direct incorporation of Me from methionine into the central tyrosine fragment (i.e., into a simple phenolic ring) and the first demonstration of its enantiospecific direct introduction into a deoxysugar. The terpene unit, puzzlingly, is not labeled by mevalonate. An acylaminocoumarin (based on tyrosine in both portions of this molecule and initially lacking the nuclear Me) is incorporated without degradation to produce complete (methylated) novobiocin. This result indicates the stage of C methylation.

A synthetic chemical pharmacologist would try to make new antibiotics by inserting halogen or other alkyl groups. Similarly, we first attempted to make **87** (R = Cl, Et, etc.) by feeding the appropriate synthetic substituted acylaminocoumarin, which was kindly supplied by Bruno Vaterlaus (F. W. Hoffmann LaRoche). We were thus effectively using a mutant as a reagent to complete the chemically difficult final steps of making and adding the sugar to form the antibiotic. We were

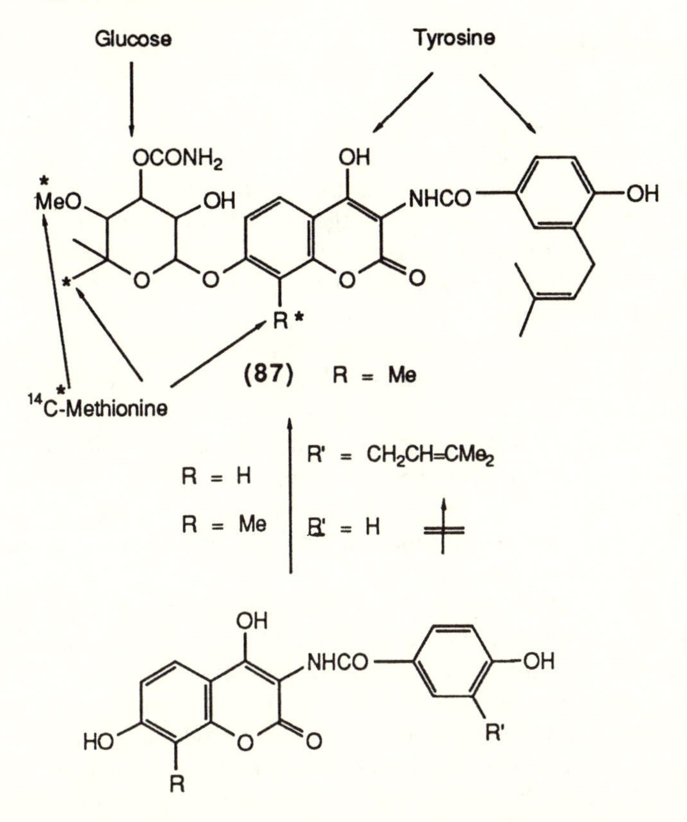

successful with **87** (R = Cl), which is a good antibiotic with rather a different spectrum of activity. With R = Et, Pr activity disappeared.

The knowledge that arose from our initial ideas and incorporations[92] confirmed that the nuclear Me, as well as the NMe_2 of the tetracycline antibiotic Terramycin (oxytetracycline), comes from methionine, and much of the rest comes from acetate. Thus we expected that mutation could lead to a desmethyltetracycline if the C-methylation enzyme necessarily involved could be destroyed. This expectation assumed that the enzymes responsible for further transformations could still deal with this altered des-Me series of substrates. Had the C-Me been derived biosynthetically by incorporation of propionate (we showed it was not), such a result would be much less likely because highly specific alteration of skeletal enzymes, rather than disruption, would have been needed. I discussed this possibility in industrial lectures in the United States in 1954. Industrial work later led to production of the desmethyltetracycline series and defined the detailed intermediate pathway to the tetracycline nucleus itself.

Chemical Phylogeny

The structures of secondary plant and mold metabolites chemically mark the diversity of species. I thought, therefore, that their structural modifications in related organisms might be useful in evolutionary classification (phylogeny) by reflecting changes in pathways due to evolutionary mutations. This approach is related to taxonomy, an artificial classification of useful arbitrary boxes into which to file different organisms, but it is more fundamental.

Molecular structures of plant components had been examined to aid taxonomy in the same way as visible characteristics like leaf venation and shape, flower and seed types, and so forth. I felt that phylogeny, concerned with evolutionary sequences, is more clearly related to predictable biogenetic changes. This approach required that structures as such should not be regarded in isolation as markers. Rather, the chemically definable evolutionary alterations of biogenetic routes to the related constituents are important.

These changes could be guessed by examining the structures in a related biogenetic series in the light of the increasing knowledge we were gaining on origins and reasons for existence, especially if one target molecular structure can be defined. For example, if all plants need gibberellins, a route to form them must exist, which in outline we had defined. If mutations of the route occur, plants can survive because of gene duplication, but they could also accumulate new diterpenoid congeners of gibberellins. Thus I explained to my own satisfaction why almost all plant diterpenes belong to one of two types: some can be defined on paper as mutated offshoots of routes to gibberellins[104] and others as derivatives of the alternative macrocyclic diterpene neocembrene (which we first isolated). Other types are very rare. One that we structurally investigated,[121,122] eremolactone, is probably basically a sesquiterpene with an introduced C_5 unit. This diterpene structure situation contrasts with the many biogenetic and interesting types of sesquiterpene skeleton, despite their five fewer carbons.

This viewpoint renders possible, with structures and general theory, to determine what steps have changed through mutation processes in the stages to such definable target molecules (ring closures and stereochemistry altered, functional groups inserted or omitted, the courses of carbenium ion migrations altered, and so forth). Such new compounds are presumably left behind as residual metabolites. Existing enzymes presumably cannot transform them, although sometimes further new metabolic transformations through new enzymes may have been induced by their presence. The products isolated are not necessarily exact indicators of the nature of a mutation step. Interpretation is required in the light of chemical theory and of biogenetic knowledge. This uncertainty is part of the fun. The formation of such secondary products may continue because they turn out to be useful, like the pine resins, as wound-healing agents, insect repellents, and so forth.

Gibberellins are not the only class of plant target molecules that can be considered in this fashion; triterpenes, flavonoids, and anthocyanins are others that may have definable functions. Loss mutations are easier than gain mutations because they represent the disruption of an enzyme. Altered pathways must be considered with that fact in mind.

In discussing the topic, I tried to estimate whether chemical changes are trivial or nontrivial, according to their chemical difficulty. Such estimates can never be quantitatively scientific, but nevertheless are not meaningless. This type of subjective approach to correlations leading to further real work is becoming increasingly respectable. Evolutionary structural changes in genes, peptides, and enzymes are more fundamental because changed metabolites are mere reflections of these alterations. My simple-minded organic chemist's approach has largely been overtaken by increasing capabilities in finding out what actually happens structurally in those fundamental areas.

Genetics of Flower Color in Dahlia variabilis

One of the entrancing features of my 1953 paper[83] was that it gave the first complete explanation of the origin through a common precursor of the skeleton of plant stilbenes and the related chalcones forming the primitive precursors of all other flavonoids and anthocyanin plant pigments. Robinson had suggested that the C_6C_3 portion of the anthocyanin skeletons, because of its oxygenated substitution patterns, was related to tyrosine or 3,4-dihydroxyphenylalanine (DOPA), but nobody had suggested with any rationality the origin of the other C_6 portion, as I did from three acetate units.

A primitive enzyme-bound polyketide-chain precursor in my view leads to the first stable isolable product as a chalcone (**88**), or by alternative ring closure as a stilbene (**89**). My suggestion for such skeletal origins was rapidly biochemically confirmed by others with better facilities for manipulation of tracers in higher plants, with the key compounds quercetin and cyanidin.[122]

One Japanese worker recorded that he used buckwheat that was "growing wildly on the campus". As a founding member of the editorial advisory boards of *Tetrahedron*, *Phytochemistry*, and the *Journal of Antibiotics*, I have, incidentally, been asked to rewrite many papers in idiomatic English. I recall writing, as a referee for a paper by my friend Kubota, "This is admirable science, but should be rewritten in idiomatic

Presenting a lecture in New York in 1954. (Photographed by Charles E. Rotkin of Photography for Industry.)

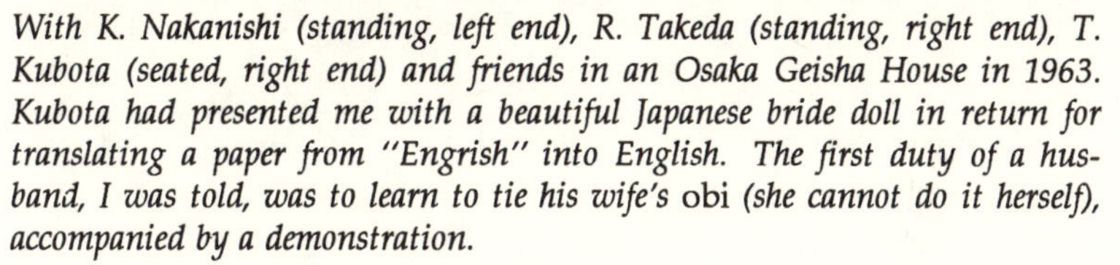

With K. Nakanishi (standing, left end), R. Takeda (standing, right end), T. Kubota (seated, right end) and friends in an Osaka Geisha House in 1963. Kubota had presented me with a beautiful Japanese bride doll in return for translating a paper from "Engrish" into English. The first duty of a husband, I was told, was to learn to tie his wife's obi *(she cannot do it herself), accompanied by a demonstration.*

English." Guess what? I soon received a letter saying, "The referee says so and so, will you do it for me?" So I did, and have in exchange a magnificent Japanese bride doll, which is still in my living room.

Solely on the basis of chemical reactivity considerations[122] about my postulated fundamental origin, I suggested that the successive steps to flavonoids and anthocyanins must be: chalcone (**88**) into flavan-4-one (**90**) (ring closure of OH onto activated double bond) into flavan-4-on-3-ol (**91**) (oxidation at an enol-activated 3-position) into flavonol like quercetin (**92**) by oxidation to a fully aromatic heterocycle. The formation of the oxonium ring of the flower-pigment anthocyanins of type **94** must at some stage involve the reduction of a 4-carbonyl, marking the initial chalcone origin, into CHOH to permit mechanistically (through a carbenium cation) the removal of the 4-hydroxyl anion. On this score I was initially in favor of an intermediate flavan-3,4-diol (type **15**), which we synthesized in 1953. However, we were misled for many years by some incorrect biochemical evidence reported from New Zealand against this structural type as an intermediate. It now seems to be correct. To summarize: Organic reactivity considerations based on a precursor and a final product successfully predicted genuine intermediates.

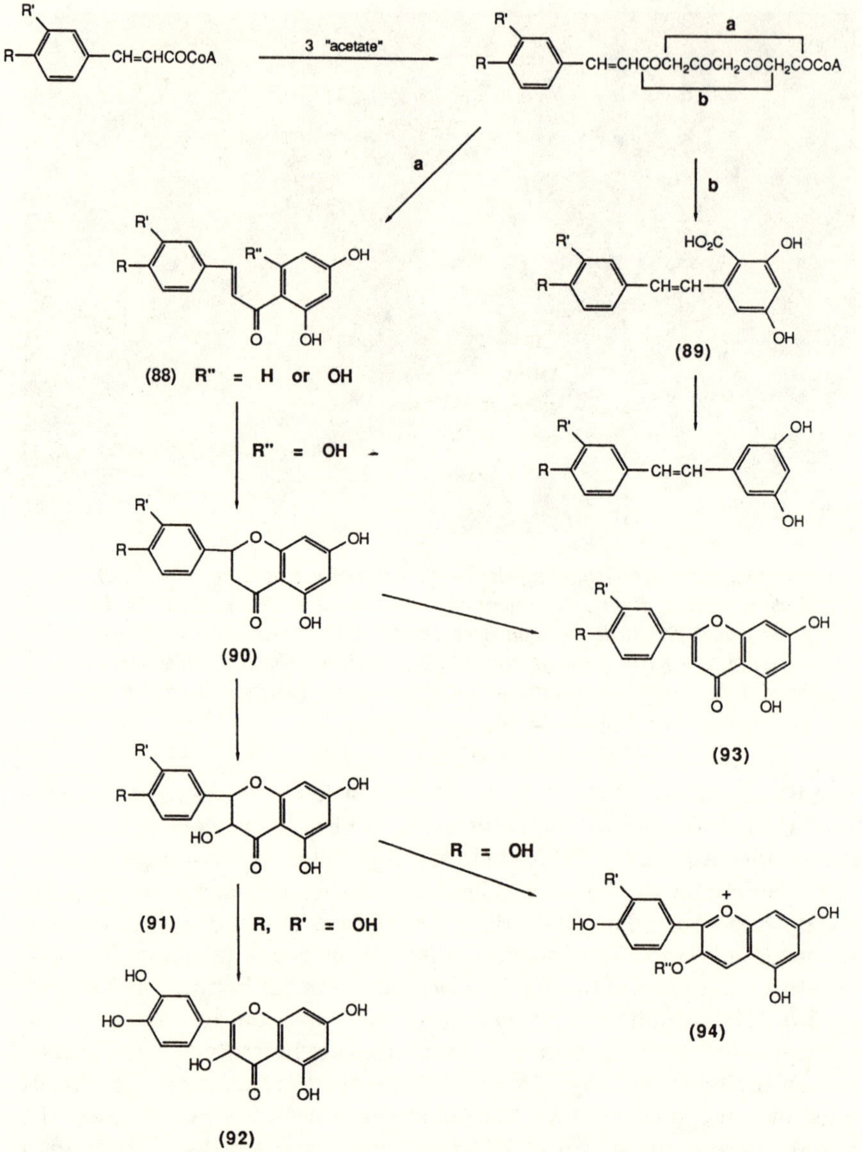

3 "acetate"

(88) R" = H or OH

R" = OH

(89)

(90)

(93)

(91)

R = OH

R, R' = OH

(94)

(92)

The biochemists as usual took no notice, some even correcting their erroneous views to my initial ones as if to a new viewpoint. After many years of biochemical fumbling, these chemically expected routes (the major ones of which I have discussed in lectures since 1953) have been confirmed. It still appears that the only respectable and accepted approach must first be approved by those within the inner circle and be the result of biological experiment. It is a pity, because speculation of my type is not only designed to satisfy the speculator, but to short-circuit experimental labor by suggesting the right experiments first (as happened initially, for example, with A. C. Neish in quercetin biosynthesis). There is far too much to do in an enormous cooperative biochemical venture to waste energies and resources for personal reasons.

Given the availability of the main series, alternative side paths to known natural series by chemically acceptable processes immediately became apparent. These side paths involve oxidation processes of several types: chalcones into natural aurones, by removal of 2H from a flavan-4-one (**90**) into a natural flavone (**93**), and by conversion of a flavanone into an isoflavone by predictable Wagner–Meerwein carbenium ion migration of aryl from the 2- to the 3-position.

Chemical Genetics

The genetics of flower color in *Dahlia variabilis* is the classical J. B. S. Haldane work on the first chemical genetics. It was not possible for geneticists to interpret their genetics in the absence of chemical models (e.g., reference 123), but I was able to do the reverse. I had the temerity, with no biological training, to assume that I could understand enough genetics. My approach moved from the fundamental to the general. I found, in Manchester, that it was possible for fundamentally trained chemists to learn biochemical techniques and ideas easily; biologists often had great difficulties with handling chemistry like chemists.

Dahlia is genetically a tetraploid, and it can carry up to four genes for any pigmentation characteristic. I started by assuming that the enzymes produced for any given step are quantitatively determined by the genetic dosage present for that step. There are four recognizable color genes: **Y** (yellow) produces chalcone (**88**, R, R′ = OH, R″ = H), which significantly has lost one of the aromatic nuclear acetate oxygens, presumably by carbonyl reduction in a ring-open β-polyketo CoA ester precursor. The process is undoubtedly controlled by **Y**. This desoxy structure (R′ = H) predictably renders the chalcone mechanistically less prone to ring closure to a flavanone, which may be why it accumulates. Gene **I** (ivory) codes for flavone (like **93**, R, R′ = OH), representing from the same considerations a genetic side-track oxidation of flavanone (**90**).

Very significantly, one or other of genes **A** or **B** must be present for any anthocyanin production. Each gene apparently controls the same chemistry, but they are different quantitatively; **B** is more effective than **A**. But they do not structurally define the anthocyanin (pelargonidin, **94**, R′ = H; cyanidin, **94**, R′ = OH).

The chemical keys to the situation are as follows. Lawrence and Price,[123] following earlier genetic work of Scott-Moncreiff and others, showed that the increasing dosage of different genes was competitive among each other for pigment formation; that is, some precursor must be limited in availability (in my view the hydroxycinnamoyl coenzyme-A series).

Although gene **A** or **B*** must be present for anthocyanin formation, they are not specific genes for the mono- or dihydroxy B-ring substitution pattern of pelargonidin (**94**, R′ = H) or cyanidin (**94**, R′ = OH), as happens genetically in some other flowers. Which anthocyanin pigment is produced depends on the total dosage of color genes for all pigments of any flavonoid or anthocyanin type. The higher the total genetic dosage, the greater the proportion of the ring-B monohydroxy pelargonidin (**94**, R′ = H) compared with the ring-B dihydroxyanthocyanin cyanidin (**94**, R′ = OH). This was the puzzle.

My interpretation[122] was to assume that the transforming enzymes present leading to anthocyanin can deal with both series of hydroxylated precursors (because both are transformed biochemically), but prefer the dihydroxy series. Therefore, so long as dihydroxy precursor is present within the metabolism, cyanidin results. However, the direction of hydroxylation in the general metabolism is normally the conversion of the 4-monohydroxy into the 3,4-dihydroxy series (tyrosine into its 3,4-dihydroxy analogue DOPA, for example). Here it is presumably controlled by a (constant-rate) general hydroxylation mechanism not tied to the pigment genes except perhaps by feedback mechanisms. As pigment enzymes are increased with genetic dosage, rate of withdrawal of precursors will eventually overwhelm this constant dihydroxylation rate. The pigment product will correspond increasingly to the conversion of the monohydroxy precursor, as pelargonidin(4-hydroxy-), despite the lower affinity of this overall series for conversion enzymes.

Predictions can be made. For example, chalcones corresponding to both series are known from *Dahlia*, but not from which genotypes they are derived. A prediction is that increasing **Y** genes should result in more monohydroxylated chalcone. I tried to test this. In 1951 I wrote to the John Innes Institute, from Cambridge, only to be told that all of the original tubers had just been destroyed. Nobody with horticultural expertise has since been interested enough to test my ideas. In the meantime, pigment studies based on genes have made great progress and effectively confirm my original anthocyanin intuitions.

---

*These genes are not related to the structural A and B aromatic rings of the flavonoid nucleus as shown; the B-ring structurally is the one that "dangles". This is one more example of the unsatisfactory, unconcerted nature of much scientific nomenclature, understandable only to "experts" in a small field.

Political and Other Activities

A Scientist in Blunderland

My obligations as a professor, especially in Manchester (1956–1967) and Canberra (1967–1980), led to a wide variety of interesting but time-consuming activities of a political nature (using that term loosely) within, but often outside, the university sphere. An appropriate question is whether scientific training fits or unfits a person for making political decisions. The answer, as usual, depends on the type of person. Broadly, it could be inhibitory because scientists tend to want "all" of the evidence before making a decision, whereas any rapid political decision may be better than none at all. However, training in rational ways of thinking and approaches to testing results can lead, in some scientists, to high political ability. Consider, for example, the Lords Todd, Zuckerman, and Dainton in the United Kingdom.

There are two major differences between politics and science, although they may start in the same way with a favored idea: Politicians strive to support their idea selectively by any means available, whereas real scientists skeptically test theirs to destruction if necessary. Admittedly, there are major difficulties with social control experiments in politics, but the attitudes are very different. The reward systems are also different: Politicians seek money and power, from which reputation flows; scientists seek reputation, from which power and money may flow, although these are not the primary considerations. The two attitudes sit uneasily together. I, for example, greatly disliked providing ideas and writing material that others signed, or for which they took the credit. In public service people are paid for that; I was not. Many

Receiving the highest Australian civil award, the Companion of the Order of Australia, from Governor General Sir Ninian Stephen.

successful scientific advisers tend to tell the politicians what they want to hear, on the grounds of realism ("if you want anything done ..."). They probably have a point; I finished in the political wilderness.

I might have avoided some of these activities, had I been more strong-minded. Bob Woodward told me that he refused to accept any positions in professional societies and refused to referee scientific papers on the grounds that many people could do that, but only he could do his research. I protested: "But Bob, you expect your papers to be refereed!" With his characteristic wry smile, which he could assume, he said, "But that is not necessary." He was probably right, but I felt obligations to the world that supported me.

Policy Definitions

Politically I was, as usual, thrown into the deep end of the pool. Without desiring it I somehow acquired power and influence. I was never taught political skills or administrative methods, so I had to invent my own often unconventional approaches. They were intended to be scientifically rational when all relevant factors are taken into account. The term "rational" as used here includes dealing with what makes people tick, no matter how irrational that may be, scientifically speaking. My plan of action was to set up a grid of principles and to operate pragmatically within it, with appropriately chosen collaborators. It was much like chemical research, but with less certain results. My success in the area seemed to be much greater than if I had been conventionally taught such skills. The only high-level course on administration I ever

At work in the Research School of Chemistry, about 1970.

attended was one that I taught. On a personal level, I was never good on committees unless I was chairing them, which was often the case. Jack Lewis (now Lord Lewis, Head of Robinson College in Cambridge) was very antiestablishment in Manchester. Now, as I pointed out to him, he is the establishment. That is the fate of all but the wary (like me).

I had various onerous organizational duties, including a repeat performance as Dean of the Research School of Chemistry (RSC) (1973–1976). I was first the Treasurer and later the President of the (independent) Australian Academy of Science. This unpaid post (4 years), which I took over at a period of its low financial ebb, involved reorganizing the whole administration (at personal costs in unpleasantness). Although the funds were yet to be found, we refurbished an old government hostel at great financial cost; it is now a very elegant headquarters. I was elected Vice President and President of the Royal Australian Chemical Institute, one of the largest professional bodies in Australia. I chose to visit many of their outposts.

I was organizer of the science part of "Festival Australia 75". Later I served as a member of a committee that successfully led to the foundation of a new "hands-on", "minds-on" public Science Center in Canberra, after contributing to a session entitled "The Science Centre, a Challenge to the Traditional Museum" in a UNESCO conference. This session offered the vision of a hands-on, minds-on approach through which the general public could understand and feel the unseen beauties of science and its methods. It would highlight the contribution of science to human civilization, including communications, social organization, and medicine, and would use any techniques that might be necessary to convey these concepts. The Science Center still has not been implemented fully in Australia.

Science and Technology for Development

On the international level, I organized a program in the nongovernmental section of the UN Science and Technology for Development meeting in Vienna (1979). I personally represented Australia at my own expense. The goal was to show how Australia, as an affluent but developing country, could define its own problems if properly viewed. (The woes of Calcutta seemed incongruous under the crystal chandeliers and marble halls of Maria Theresia's Hofburg).

I was, in addition, an examiner for the Organization for Economic Cooperation and Development (OECD) in Denmark (Science and Technology Policy). Denmark has problems that are very similar to those of Australia. The situation offers lessons to be learned, but Australian politicians do not want to know about it. As a consequence, they continue to reinvent the wheel. In six visits to the People's Republic of China I helped to organize and fund research under the UN Development Programme. More importantly, I was there to advise at political levels on how to set up technical and laboratory management and instrument centers. With the support of others (notably my laboratory manager John Harper) this activity has been, I believe, very successful in improving the future of a quarter of mankind, a great reward in itself.

Advice within Australia

Within Australia, as Chairman of the Science Policy Committee, Science and Industry Forum, Academy of Science, following my predecessor Sir Geoffrey Badger as advocate, I helped in 1970 to set the blueprint of what later became the Australian Science and Technology Council. I was briefly a member of the first version (1976–1977), which was dismissed by a new government. Later (1983–1985) I prompted the Academy to catalyze—financially, ideologically, and organizationally—the formation of the Federation of Australian Scientific Societies (as a lobby group) and the National Science and Technology Advisory Group (to comment annually on government budgets). Those operations had to be handled very carefully. Professional societies viewed the academy as an elitist body so they needed to be led to think that they had come up with the ideas themselves.

I was asked by Prime Minister Malcolm Fraser (I believe on the basis of my AAS Forum activities) to chair a small committee (1976–1977) commissioned to define what the Commonwealth Scientific and Industrial Organisation should be doing. This important task was concerned with the largest and broadest operative Australian government research body, which had not been examined for 30 years. I asked Fraser if there were any political aspects that were not specified in the terms of reference. "No, I just thought they were getting as bit slack!" We reported our conclusions in 1977 and most of our recommendations were accepted, although a few important ones were not. These government failures and those of the CSIRO scientists, who refused to recognize that they had been served notice to be realistic, led later to another inquiry. Our major conclusions were reaffirmed, but CSIRO was placed firmly under the control of the politicians, not the scientists and technologists.

I was the founding Chairman of the Australian Marine Sciences and Technologies Advisory Committee and its funding panel. When the Prime Minister asked me, I objected that I am not a marine scientist. "That is the reason." I soon appreciated his realism. I had advisers who could see tides, whales, ocean floors up to a certain depth, the Great Barrier Reef, ocean currents, and so forth, but I was the scientifically trained "ignoramus" who had to ask the broad *obvious* questions for the good of Australia as a whole.

An amusing job was chairing an inquiry into further and continuing education; it introduced me to social science jargon. When I asked for written elucidations of terms, I received another set of equally opaque social science jargon. I soon realized that science jargon is also obscure to the uninitiated, and have since tended within that context to use circumlocutary explanations rather than scientific words.

What Classes of Research?

In the course of the CSIRO inquiry and later my marine science committee work, I had to formulate and answer some very fundamental questions about classifications and objectives of research and its funding. In 1976 I rejected existing divisions of research into pure, applied, long-term, and short-term. Instead, I reclassified research in terms of intentions and technical types: strategic (divided into uncommitted and mission-oriented) and tactical (problem-solving). All research is pure; similarly, all is applied in different proportions over different time scales. We cannot define most situations in advance. The techniques and rationale are the same for each class.

During a trip around the world at the time of the CSIRO inquiry (1976) I discussed the ideas widely with science-policy people abroad, and there was some international reaction. The nomenclature reached by OECD at a meeting in Frascati adopted my classification in part, but set up a series of headings on "socioeconomic" objectives, which blurred my rationality. Is all university research only to promote learning? Is all medical research different from the science on which it is inevitably based? Does not "applied" research also contribute to fundamental understanding? Does not "pure" research form the inescapable basis for applications? The problem with government funding results from separate "buckets" of research that do not interconnect and are labeled so that politicians who do not understand the subtleties can inappropriately pour money into one or another. An international problem is that research comparisons are made on the basis of such labels.

Because statistics are now largely Frascati-based, international comparisons are more difficult, as I see it, than they should be. However rational it is in connection with how research really works in historical and practical terms, my proposal has been too unconventional to grasp for most politicians and public servants and even for some scientists, notably those within the science policy area. Scientific–technical problems are resolved in the end not by human votes, which can be organized and manipulated in the short term, or by political dictates, but by the long-term impersonal verdict of Nature.

These activities took 6–8 years away from my own research, at a period when it could have been most fruitful because of my available support facilities. The academic peak of my recognition as definer of policies was, I suppose, when I became the subject of a successful Ph.D. thesis in political science at the University of Melbourne by Dr. Joan Clarke, now executive head of a very successful mental health charity.

A Commonwealth Minister once said to me, "I know all about universities, I have a Ph.D. in political science." "Indeed, Mr. Minister, but I am the subject and supervisor of one, and what is more, I passed." (That was how I gained friends among politicians!)

Rewards

A friend, commenting on his British knighthood, said to me, "This is what the politicians give me instead of paying me." In Australia one is given neither of these rewards, but there were others. Such activities were personally very flattering and emotionally rewarding to my ego, not least because I believe that I was able to exemplify the applicability of scientific rational thinking outside classical science. I did derive some technical research advantages from this kind of conduct. For instance, after frequent short interruptions I returned anew to each scientific problem. This renewal broke vicious circles of thinking, which are the greatest hazards of those scientists who are permitted to revolve in continuously decreasing mental circles within their laboratories. Often their thinking slowly becomes introverted, outdated, and irrelevant.

My political reward was Companion of the Order of Australia, which in the Queen's "pecking order" is higher than most knighthoods, and rather equivalent to an English Lord, but carries no title. It is perhaps equivalent to Chevalier de la Légion d'Honneur in France.

Research School of Chemistry, Australian National University (1967–1980): A Unique Opportunity

A major involvement from 1967 to 1980 in jointly establishing a new Research School in Canberra along entirely new lines tended to inhibit my personal research. Later this handicap was partly offset by the lack of undergraduate teaching responsibilities. (I finally left Manchester in 1967; my work there included teaching many courses, including elementary ones.) Few people have an opportunity, in their 50s, to organize their dreams into novel practical realities. This chance was given to me and David Craig when we were asked to set up a Research School of Chemistry in the Institute of Advanced Studies of the Australian National University. I was the foundation Dean (Director) from 1965 to 1970 and again from 1971 to 1975. Of course, the onerous organizing duties involved in setting up an entirely new scheme, together with responsibilities within the overall university system, detracted for at least 8 years from my own researches. But the responsibilities can be joyous ones, with the pleasure of beating down hide-bound administrators.

Our operation was unique in several respects. We could we set up a novel organization, unprejudiced by history (except for some general ANU rules about staff classifications). Moreover, our particular circumstances gave us an unusually free hand to do what we could see was obviously needed to set up a world-class research organization that would operate on novel science-based lines in a rather remote country. Many previous Australian professors had been "pommies" (English); they lacked a due understanding that there is a wider world and that Australians are not an English subgroup with a strange accent.

We were able to accomplish this task because the ANU and the Australian government were both eager for its success. However, the government imposed one condition; the project could not go ahead unless two out of three invited expatriate Australians agreed to return. The expatriates in question were Ronald Nyholm, David Craig, and me. We were all in very good situations abroad (we occupied academic Chairs at major universities). At one time or another we each refused to return to Australia. Nyholm finally decided to stay in England, but not before he had had considerable influence on the developing project. The government, through Prime Minister Menzies, offered the ANU (if we were involved) about 1.25 million pounds (2.5 million American dollars) in addition to any Australian Universities Commission funding. This amount was enough to construct and largely to equip the building, equivalent to perhaps 25 million American dollars or more today. It was said then that we were worth more than football players!

David and I were consequently in a very strong personal position when we did return. We were able to set up our own novel organizations in ways designed to meet long-term, rational requirements that were obvious to us as research scientists who had fortuitously been little prejudiced by indoctrination into the standard university organization. We desired to achieve a valid steady-state situation for the fore-

Sir Malcolm Fraser speaking at the opening of the Research School of Chemistry. Listening on the platform were (from left) H. C. Coombs, Lord Alexander Todd, and I.

seeable future, in relation to scientific opportunities and requirements both for obtaining results and for training people.

Organizational Features

The school we organized differered from the others already existing, or since set up, within the ANU Institute of Advanced Studies. The traditional schools are fragmented into departments with earmarked staffing and budgets. Our operative objectives were continuing flexibility of building and of staffing that allowed us to respond to perceived changing scientific opportunities and the appearance of first-rate researchers in any area of chemistry, together with strong and virtually automatic updating of instrumental and technical support. "To each generation its chance", "unlimited advice, unlimited explanation, and uninhibited final decision" (by the Dean) were some of my initial dicta as the Dean-elect. This approach required from the two professors a considerable degree of personal unselfishness and dedication to the chemistry of the future. We could easily have arranged matters more for our own short-term benefit, as had mostly happened elsewhere within the ANU.

The School was to be, and is, an overall center of excellence in its facilities. Specific research programs operate as a set of changing individual centers of excellence, each clustered around a few excellent tenured staff members and selected research problems. The organizational staff and budgetary flexibility allow the school to tackle new objectives and incorporate new excellent people as they become available. This organization required two sets of staff, both excellent but in different ways: some are concerned with the overall center (chiefly in instrumental areas), and others who head flexible groups of research excellence.

In Australia, the opportunity to obtain research funds from outside the university has been extremely limited, particularly for the Institute of Advanced Studies, ANU. Thus our overall flexibility was largely dependent on how persuasive we were with the ANU funding. Nevertheless, we must be grateful in many ways for that ANU financial organization. I recall Bob Woodward telling me that his grant from Harvard paid half the salary of his secretary. He had no continuing funds for a microanalyst, who spanned many individual research projects. There is a desirable mean.

Our organizational schemes were all worked out in the ways that scientific problems would be attacked. We incorporated the additional human, financial, psychological, and political factors with the scientific ones, against a background of how creativity has to be promoted. As in scientific research, reactions to evolutionary changes were assessed as the results of applications in practice became available. Our theory relied on a number of organizational features then unknown on the Australian scene and in the ANU (or elsewhere, so far as I knew). We established a high nontenured staff ratio; no departments were separated (because chemistry is a unity), but a series of changing research groups; and a strong but rotating Dean as an advised dictator for a limited period, rather than a long-term Director as found in other ANU schools. New administrative methods had to be devised to cope with a situation in which a high proportion of nontenured staff posts is continuously available and all of the central research funds need to be continuously reallocated each year on the basis of requirements that are expressed on paper in carefully considered ways. (Experience has demonstrated the validity of these approaches.)

The system we evolved, which seemed obvious to us as working scientists, was often abhorrent to administrators and even to many classical academic staff members with inferiority complexes. Part of the problem was that our system put decision-making within the School, in the hands of those most technically qualified to exercise it. This structure removed power from outside: from allegedly unbiased (but ignorant), referees from the administration, except for consideration of the overall School budget. Our system did not fence off departmental structures from the needs of the School as a whole. Because we had none, departments did not have to be approved or their Heads appointed by outside action. We could respond rapidly to new opportunities of any kind, including the abolition of existing activities.

Our new organization deliberately, continuously put the onus on the faculty individually to justify their work. I think the university did not believe that scientific calculations could set up the novel methods needed to cope with the selfish demands of a lot of competing researchers. On the contrary, nowadays the ANU (seeing that the School is successful and as flexible as the day it was founded 27 years ago, even in the face of contracting funding) is beginning to imitate our organization elsewhere in the Institute of Advanced Studies. I confidently defer to history on the success of our human–scientific experiment.

A Novel Building

Even our approach to the building was novel. We employed architects like Rod MacDonald and Bill Batt, who tried to find out what we really wanted rather than insisting on unscientific "architectural features" such as are seen in some other ANU schools. The design had to be flexible; there are no internal load-bearing walls, unlike the usual boxes designed by people who did not know what they really needed, with many leaving before completion (as in the later unduly expensive Research School of Biological Sciences, ANU).

The structure was largely prefabricated to a single "module" for easy alteration, which has happened extensively. It was, unusually, finished on time at projected cost. It was very consciously designed for scientists by scientists with sympathetic architects. It reflects our whole attitude to the new School: not just for ourselves, but for the indefinite future. It has in 27 years proved to be as flexible as we planned. It is beautiful and, because of the functional design, economical to run.

All of this planning depended on a fundamental knowledge of needs in practice and how these needs may change, even as a result of techniques and topics not yet discovered. This was combined with the "political" power to express our opinions. We were fortunate in having many ANU administrators such as Ross Hohnen (ANU Secretary) who backed our judgments. We deliberately situated our building next to the undergraduate school, with which we interact, rather than physically with the Institute of Advanced Studies. This decision was reached in spite of considerable opposition, which has recently been revived by idiot politicians.

Research, 1965–1974

My biosynthesis area was winding down after 1967. It climaxed with stable isotope work by Tom Simpson on phomazarin (Research Set 3).

The Research School of Chemistry, 1972.

Tom has since gone on to new heights in the general area. Some total-synthetic aspects based on "Birch-reduction" dienes were continued, notably the development of Diels–Alder and Alder–Rickert reactions (Research Set 2). This work led to my gradually increased appreciation of the principle that I later (in discussion with Richard Stephenson) styled "lateral control" of structure in laboratory synthesis, although for the most part I developed the concept experimentally not only through my applications of the Diels–Alder reactions but also later through transition metal complexation.

I continued to use the Birch reduction products of substituted aromatics (Research Set 2) as experimental vehicles for much of my synthetic work of all types. A very large variety of pure compounds is available with definitely positioned substituents and unsaturation, including unique structures such as regiospecific enol ethers in unconjugated cyclohexadienes. And we knew how to make them.

From about 1962, but mainly throughout 1967–1985, I undertook a broad case study in organic synthesis using organometallic compounds. The very laborious and systematic study was attacked from the point of view of the typical organic chemist, on the principle of the need of at least one functionally multisubstituted metal series. In my chosen case the experimental work was focused on $Fe(CO)_3$ and cyclohexadienes (for motivation and details, see Research Set 4).

We obviously could not neglect known organometallic theory and practice. However, we placed a new emphasis on aspects not normally covered by organometallic chemists of the time in their concentration on the transition metal atom and the nature of its bonding with simple, often unsubstituted, organic substrates. We were closely interested in the effects of complexation, not only on the reactivity of the functionalized organic portion itself but on functional substituents, as outlined in Research Set 4.

Unfortunately for me personally, apart from a few examples like gabaculine and isotopically labeled shikimic acid (Research Set 4), lack of practical research support in my "retirement" has necessitated leaving exploitation to others, notably to my very able former collaborators, A. J. Pearson and Richard Stephenson.

Research Set 4

Lateral Control of Synthesis: Transition Metal Complexes

Classical organic synthesis depends largely on what I have named "endogenous control": bond-forming reactions mediated by functional groups that are part of a main molecular skeleton or attached to it by organic bonding. Steric control is to some extent exercised by conformations of molecules and transition states of reactions, as in the Diels–Alder series discussed in Research Set 2. For complete and efficient steric and enantiomeric control, what I name "superimposed lateral control", such as that used by enzymes, is required. In principle there are several ways to exercise this control in the laboratory. One method involves extension of the elements that are normally used synthetically to the whole of the periodic table. Transition metals are especially included because of their geometrically directed bonds and their ability to produce on demand bond-forming electronic activations (reception or donation) through their d-orbital electrons.

My vision was to surpass in the laboratory the capabilities of enzymes through unlimited access to unbiochemical types of reactions and through using reaction conditions and reagents that are not compatible with the structural features of natural enzymes or even of their genetically engineered extensions. My philosophy is to use the principles by which enzymes work, without trying to imitate the exact practical methods.

Enzymes selectively assemble molecules (often chemically activated by combination with coenzymes), arrange them structurally, and promote, sterically and chirally, bond-forming processes within an asymmetric molecularly engineered cavity. Instead, we try to achieve similar assembly and activation, including intervention of the equivalent of coenzymes, but around a geometrically defined point (a transition metal atom) instead of within an atomically "engineered" hole. The synthetic results can in principle be similar, or identical, especially when we are using optically resolved reagents or complexes.

The Creation Center

My initial small creation center for this work was my need for pure conjugated 1-methoxycyclohexa-1,3-dienes for synthetic reasons (Research Set 2). I hoped to reach them through $Fe(CO)_3$ 1,3-diene complexes[124] derived directly from unconjugated "reduction" 1,4-dienes. The isomers **95** (R = OMe) and **96** were both found to be obtainable from the unconjugated 1-methoxycyclohexa-1,4-diene by complexation with $Fe(CO)_5$. I tried to separate the isomers through hydride abstraction with trityl fluoroborate to give crystallizable "carbenium" salts, a known type of reaction with the unsubstituted complex **95** (R = H), hoping that the pure isomers could be regenerated by reduction. The stable salt **97** was obtained together with, unexpectedly, the neutral dienone complex **98** clearly resulting from hydrolysis of a 1-OMe "oxonium" precursor, a fruitful accident.

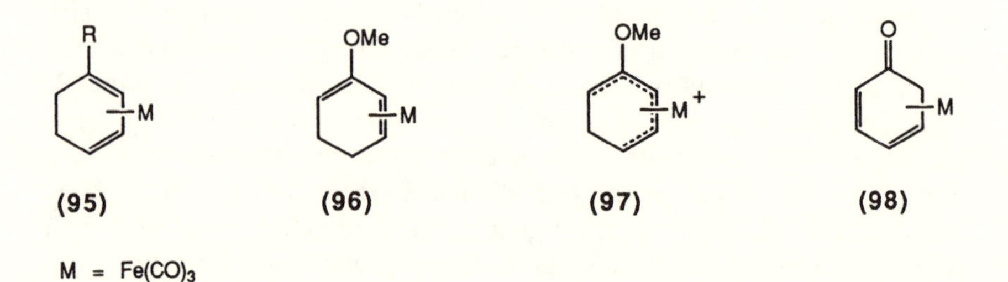

I immediately realized that both of these types of structure had further synthetic organic potential through the principle of lateral control of structure and reactivity conferred by the $Fe(CO)_3$: the salt as a sterically directed stabilized mesomeric, potentially asymmetric, carbenium ion, and the dienone as a similarly sterically directed stabilized ketonic tautomer of phenol (showing some carbonyl reactivity). My original limited aim of making specific dienes, although partly achieved by reduction of **97**, was rapidly replaced by other much more important creation centers. We used, first on paper, a series of substituted cyclohexadiene complexes and appropriate reagents, which we explored experimentally as synthetic organic (distinct from organometallic) reagents.

The following characteristics were rapidly apparent to us. The $Fe(CO)_3$ group almost always (there are minor explicable exceptions) distinguishes in reactivity 100.00% between one face of a molecule and the other. This situation is almost impossible to achieve with solely organic molecules. Reactivity here can be on the same or on the opposite face, according to the mechanism (electrophilic or nucleophilic).

Unsymmetrical organic molecules, such as those derived from 1-methoxycyclohexa-1,3-dienes [e.g., **95** (R = OMe) or **96** stably bound to such a metal atom], possess chirality. My recognition of this possibility came from a partly remembered discussion that I had many years ago in Canberra with Sandy Ogston (Oxford) about a similar situation that he defined for enzyme-catalyzed processes. By conversion into organic products of known absolute configuration (e.g., **97** is the enantiomer of **98**, R = H) into natural cryptone (**99**, R = H, R′ = isopropyl), we can deduce those of the complexes, and therefore the absolute configurations of any new asymmetric center resulting from further conversions with any reagent. For example, the synthetic equivalent of the enantiomer **99** is **102**.

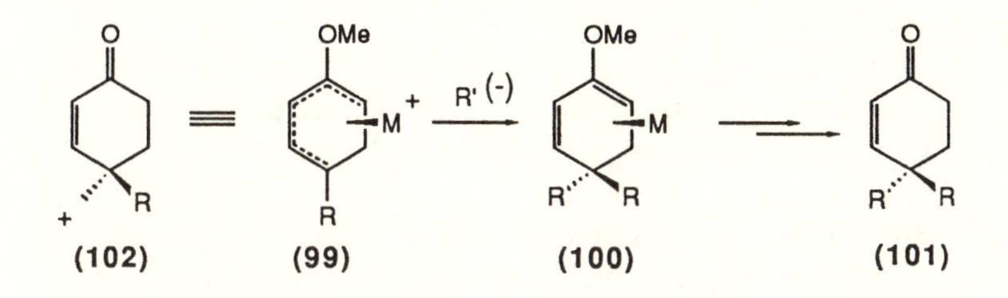

Superimposition of $Fe(CO)_3$ results in new types of activity for synthetic purposes, notably otherwise unachievable C–C formations by action of nucleophiles as shown for the conversion of **99** into **100** and then into **101**. Complexation is activated for the type of molecular junction I considered desirable under the convergent synthesis philosophy in Research Set 2 by replacing the standard organic activating groups with "molecular scaffolding" that is easily attached laterally and easily removed, here by very mild oxidative reagents. Other organic synthetic uses of the complexed organic molecule are also observed. A notable example is stereo- and enantiospecific attachment to the complexed carbon skeleton on the same face as the metal of cationic groups like acetyl (Friedel–Crafts) or protons (or deuterons). I shall discuss uses for this reaction later. Isomerizations of initially formed complexes from

"Birch" unconjugated dienes also produce a definable series of isomeric conjugated diene complexes, including ones that are not directly obtainable, and therefore lead to new synthetic potentials.

Adding together the ease of reactions and preparations, the complete stereospecificity (and less effectively at present the regiospecificity related to substituents), and the controllable asymmetry situation, it should theoretically be possible to generate readily in defined molecular situations, new asymmetric centers in full resolution and of known absolute configuration, if that of the complex is known. Knowledge of the resulting configuration is independent of the exact reagent, provided the steric course of the mechanism involved is known.

Although such ideas did not spring fully armed like Pallas Athene from the head of Jove, I was convinced, from about 1962, that I could open up a new era in my work and attempt to pioneer a new organic synthetic field. I had been becoming bored with the biosynthesis work, which had taken on the character of crossing "t"s to confirm my suggestions. But what was needed in the laboratory? The little work I had initially carried out showed that compounds in this series could be manipulated like organic ones, with a few extra precautions, so I did not have to learn a whole new practical approach (dry boxes, nitrogen atmospheres, etc.).

Before about 1962–1970

Organometallic work with organic compounds before about 1962 (when I really began) had several characteristics.[125] Because of industrial potential, the most advanced synthetic organic work was based on catalytic procedures with transition metals on "simple" substrates like ethylene and acetylene. Organometallic chemists then were interested mainly in organic–metal bonding, only slightly interested in the most basic aspects of the reactivity of the complexed metal–carbon system, and virtually not at all in the influences of functional substituents.

I avoided the usual approach and worked on stoichiometric reactions, except for the Wilkinson hydrogenation catalyst. But even in this case I employed, for the first time, structurally and sterically complicated organic molecules[126] or hitherto uninvestigated organic substrates of synthetic interest like type C=C in benzoquinones, halogenated compounds, S derivatives, or nitro derivatives. I demonstrated no scrambling of deuterium label, or of hydrogen, or disproportionation, in reducing dihydrobenzenes (in contrast to other solid hydrogenation catalysts).

My initial interest here was in a pure 2H (or 2D) donor reagent for synthetic purposes. This supplements synthetically my earlier metal–ammonia reagents. My reagents add electrons to unsaturated systems as the first step in forming mesomeric charged intermediates, whose reactions determine products. These sequences add the equivalent of 2H directly to a double bond.

Strategic Requirements

I reached several conclusions around 1962 for stoichiometric processes.

- I needed to use a cheap metal. I had fortunately started with Fe (which happens to be technically very good) and later I also used Cr.
- I needed to explore organic methods to make a wide series of compounds with a variety of functional groups in defined situations, to turn organometallic chemistry into "real" synthetic organic chemistry.
- As a practical synthetic organic chemist interested in yields after many stages, I had to find highly efficient methods to make and to convert the required complexes. I needed to know how to add the metal atom, how to carry out reactions on the complexed system, and finally how to remove the metal. For synthetic purposes I also needed to find out how to make fully resolved chiral complexes and how to determine their absolute configurations.
- I had to find out what effects complexation had on the classical reactivities of functional groups and what new (preferably regio- or stereospecific) C–C bond-forming processes complexation alone makes possible. I had to develop the standard reagents for conversions of functional groups to be compatible with the sensitivities of the organometallic structures.
- I had to apply new methods, preferably physical ones like NMR and MS, to define structures and configurations of precursors and products in any complicated substituted series.
- I had to define rates of reaction as controlled by substituents, particularly in reactions involving cations with varying substituents.
- I hoped also to develop some theoretical aspects of structure and reactivity in the chosen organometallic series that would be of correlative and heuristic value.

When I began this ambitious experimental program I did not realize how long it would take. Nearly 20 years passed before I got around to using the results in "real" organic syntheses. This time scale was partly a matter of choice, the result of continuing to pursue general objectives rather than specific synthetic ones. The work was highly successful in attaining most of my general objectives. The multisubstituted cyclohexa-1,3-diene-$Fe(CO)_3$ series now represents the most systematically investigated set of functionally substituted organometallic compounds).[127] In my retirement I cannot fully exploit my results. However, they lead to possible applications to specific syntheses, several of which I shall discuss later.

The synthetic applications seemed obvious to me. Nevertheless, general methods and approaches will only be fully noticed and publicly effective if they are accompanied by specific demonstrations. I was lax on this point, as usual, but I always had limited research support because of my responsibility, as the Professor, to provide it for other deserving staff members.

I have reviewed the general organometallic field in organic synthesis[125] and aspects of our own work such as preparations,[128] spectra and theoretical implications,[128] new reactions,[129] neutral nucleophiles to meet organometallic requirements,[130] organic substituent reactions,[130] Cr complexes, and some organic applications.[131] The flavor of our results can be sampled briefly through a few selected applications. Other broad aspects of lateral control can only be appreciated by reading our original papers.

Organometallic Preparations

Work in a new synthetic field requires discovery of how to make specific compounds and how to determine their detailed structures, including their steric configurations. My first vehicles, of course, were my substituted cyclohexa-1,4-dienes. These compounds were already available as large numbers of unique defined structures arising from the metal–ammonia reductions. I knew how to make them, and the uniquely facilitating presence of enol–ether groups contributed reaction "handles". Three alternative pathways from the same substituted Birch-reduction dienes were systematically evolved[127,129,132] to produce efficiently, with an iron carbonyl, specific different substituted complexes:

- direct reaction of the substituted 1,4-diene, in which the product (often mixed) depends on the organometallic "conjugation" mechanism
- preliminary conjugation of the diene, in which the product depends mostly on the thermodynamic ratio of the uncomplexed dienes obtainable

- acid-catalyzed isomerization of the complexes obtained by both methods, when the final product depends on the equilibrium ratios of the complexes themselves

These investigations produced some theoretically interesting mechanistic and thermodynamic conclusions, as well as a range of definable products from the same precursor.

Many isomeric complexes obtained in mixtures can be separated by chromatography, such as 1- and 2-OMe cyclohexa-1,3-diene-$Fe(CO)_3$ formed from 1-methoxycyclohexa-1,4-diene. Overall yields of complexes under carefully worked-out conditions[133] can be 80% or more. The more useful complexes are made as intermediates near the beginning of a sequence using readily available material with which alternative conditions can be examined. Complexes requiring very valuable polysubstituted material, appearing late in a long synthetic sequence are less desirable. However, a newly reported practical method we have not tried (involving ultrasound) may solve some of the experimental problems.

An outstandingly useful reactive entity is a complexed carbenium ion such as **99**. These substituted ions, in the $Fe(CO)_3$ series, undergo mainly regiospecific reaction with nucleophiles, usually at a terminal carbon of the unsaturated system, directed by the substitution (e.g., mainly in the 4-position to OMe) to form products similar to **100**. In contrast to the highly acidic specifically protonated benzenes to which the complexed cations are cognate, they usually undergo nucleophilic reaction on carbon rather than proton removal. Nevertheless, this deprotonation reaction can occur with some structures and some basic reagents (e.g., **103** is rather sensitive). The reactions are totally stereospecific on the face opposite to Fe with all reagents except complex hydrides.[134] In this case hydride may be added to some extent via the metal, a useful reaction to form mixtures containing otherwise-unavailable stereoisomers.

Removal of the iron from the resulting complex (best with neutral trimethylamine oxide) leaves a substituted diene. In the OMe series this compound is a 2-enol ether that is easily hydrolyzed directly into a conjugated α,β-unsaturated ketone (**101**) rather than automatically equilibrated with the β,γ-isomer (when one R = H), in contrast to the hydrolysis of enol ethers of the uncomplexed 1,4-dienes themselves. The eventual products obtained form a wide range of synthetic precursors,[128] some containing quaternary carbon atoms. The structure is determined by the substitution of the original complex, the nature of the nucleophile, and whether or not the original complex is optically resolved.

To show at one glance the organic synthetic possibilities of cations such as **99** and **103**, we have represented them as synthetic organic equivalents.[127] For example, based on their transformations into organic products as shown, **102** is the synthetic equivalent of **99** because of final transformation into **100**. Compound **103** is an equivalent of **104** because of the reactions shown; there is no classical organic way to generate such a cyclohex-2-enone 5-cation equivalent. When the cations are resolved the equivalents are asymmetric and can be of known configuration. My definition of "equivalent" here involves a sequence of not more than three experimental steps.

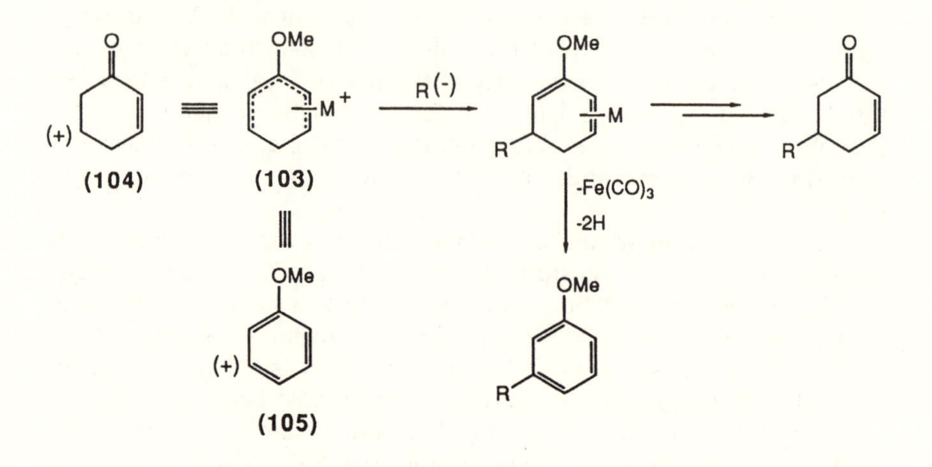

As a further example, the 3-methoxy cation (**103**) is, through dehydrogenation of products by DDQ as shown, formally equivalent to a 3-cation of a 1-methoxybenzene (**105**). Also, a benzenoid 1-cation equivalent represents the synthetic potential of cyclohexadienone complexes such as **98** because of the nucleophilic reactivity of the carbonyl and then generation of the substituted benzene by removal of the Fe, accompanied by dehydration. An experimental example of this is the direct formation of unsymmetrical substituted diphenylamines by reaction of dienone complexes with aromatic amines.[135] The neutral complexes themselves are equivalent to substituted benzenes by loss of 2H and may contain substituents introduced by their reactivities.

Alterations of Classical Reactivities of Functional Groups

To be able to use complexes in synthesis, the organometallic portion must be able to survive standard synthetic reactions on functional groups, such as lithium aluminum hydride (LAH) or Grignard reagents on carbonyls, esterification and hydrolysis, and so forth. Fortunately, the only reactions causing organic synthetic problems are oxidations, but it is even possible to oxidize a carbinol into a carbonyl with chromic acid under controlled conditions and still retain the Fe. Sometimes there are additional useful properties in altered rates of reactivity of substituents. For example, an ester group on the same face as the iron, or on the 1-position of a complexed cyclohexadiene, is inhibited for alkaline hydrolysis compared to one on the opposite face. Half-hydrolysis of appropriate diesters is thus readily achieved[130] with no particular experimental precautions, unlike the uncomplexed diesters themselves.

Complexation can be used in a number of ways to protect or to change the reactivity of dienes and groups adjacent to them. The cyclohexadiene structure in the alkaloid thebaine can, for example, be made to undergo a hitherto unknown skeletal rearrangement into a new alkaloidal series.[136] By alternative treatment, the thebaine–$Fe(CO)_3$ complex (by tieing up the methoxydiene system) can, during N-demethylation, retain its skeleton without rearrangement. The metal can be removed later to give the nor series with the original skeleton or, by further reaction, an N-substituted series with a group like benzyl replacing Me (some analogs have useful biological activities).

Some Manipulations of Structures for Organic Purposes

At least two types of rational manipulation were found to have major synthetic promise in this series. The groups attached to the metal can be altered by replacement of CO by other groups, predictably to improve electrophilic substitution of the organic system. A Friedel–Crafts type acetylation of the cyclohexadiene–$Fe(CO)_3$ complex itself, which gave low yields, was greatly improved[137] by replacing one of the electron-attracting CO groups by PPh_3 (a suggestion made by Tony Pearson).

Nonbasic nucleophiles, which also do not donate reducing electrons, had to be devised as nucleophilic reagents to form new C–C bonds efficiently with the cations. This reaction provides an experimental contrast to standard alkylating reagents like LiR or alkyl Grignard reagents, which tend to act as reducing agents or bases. We found, accidentally and surprisingly, that the use of dichloromethane as a solvent for the cation salt permitted good yields with LiR. The student who tried this solvent did not ask me first, which is just as well. Because the cations are largely "naked" ones, we thought that a variety of processes using Si or Sn derivatives rather than Li or Mg might also be favorable. Very high yields were indeed obtained with trimethylsilyl enol ethers (higher even than with enamines), and with allyltrimethylsilanes.[138] Good processes were devised for reacting the cations with formation of bonds such as C–C, C–N, C–S, and C–P. These reactions yielded a variety of novel substituted cyclohexenone, cyclohexadiene, and benzene derivatives.

Some Special Organometallic Reactions

The following reactions, which are only examples, lead to novel synthetic organic precursors or provide alternative ways to make known ones like organometallic cations. They also illuminate fundamental organometallic mechanisms. Virtually all were new when we published except the basic protonation reactions, which had not been extended fully to deuterations and their consequences.

Enantiospecific Deuteration. Protonation (or deuteration) not only isomerizes the $Fe(CO)_3$ complexes by initial addition of H or ^2H to an "end" of the complexed diene system (which end depends on substitution), but it does this entirely stereospecifically on the face occupied by Fe. In the cognate open-chain series the overall product is a cation. By contrast, a subsequent isomeric diene system is formed in the cyclic series, also by totally stereospecific emission of a proton from the same face as the Fe. No rearrangement occurs if H cannot move on that face for steric reasons. This procedure permits the introduction of deuterium in the cyclohexadiene series entirely onto the complexed molecular face (β-) exclusively in positions allylic to a cyclohexadiene system.[139] One synthetic exploitation of this is as follows.

If a complexed system is asymmetric, such stereospecific deuteration and resolution should yield theoretically deuterated products such as **108** (R = ^2H) or its enantiomer **109** (R = ^2H). The deuterated center of these products is fully resolved and is of known absolute configuration if that of the complex is known. The capability to form such asymmetric centers with deuterium versus protium resembles that of enzymes and applies to all hydrogen migration processes in the complexes.[139]

Neutral OMe Complexes into Cations. One example among many of an acid-catalyzed reaction, as a novel general procedure for making cations from available neutral OMe complexes,[140] is the conversion of the stereoisomer **106** into **107**. Like all similar acid-catalyzed reactions, this process is totally stereospecific. It results in movement of a ^2H on the same face as the metal (via the metal?) from the 4- to the 1-position. The alternative 4α-H stereoisomer of **106**, with

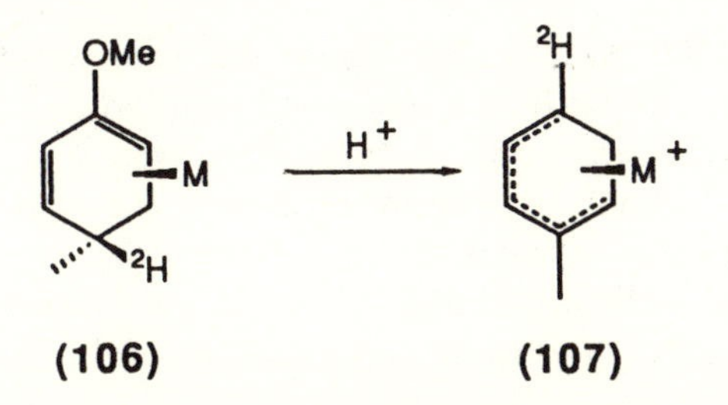

H or ^{2}H on the face opposite to the metal, does not manifest the OMe-loss reaction at all (it racemizes if optically active).

Transfer of Asymmetry. Other very interesting examples of stereospecific complexation and migration result in complete transfer of asymmetry from a CH to the molecular complex and eventually back to another distant CH. For instance, (+)-limonene (**110**) is converted[141] into **111**, the optically resolved derivative of α-terpinene. (This hydrocarbon itself is unsymmetrical but not chiral, so it cannot be a free intermediate in the conversion.) Eventually, among other derivatives, **110** is converted into the optically active phellandrene complex **115** of known absolute configuration, as shown. These terpenoid transformations shown include stereospecific and enantiospecific complexation, stereospecific acidic isomerizations with movement of H only on the metal face, and regiospecific reactions.

Formation of a substituted cyclohexadienone structure **113** from the resolved natural product carvone (**112**) is another way to make the interesting class of dienone complexes in optically active form. The absolute configuration of the complex was confirmed as that to be expected from its precursor by the sequence: reduction, treatment with acid to a cation, and regiospecific reduction into the optically active **111** further converted by acid into the known (–)-phellandrene complex **115**. The last complex can be generated predominantly by complexation from **1**, my favorite compound of 1937, together with a lesser proportion of the more hindered **114**. Only one of the two phellandrene complexes (**115**, with a 4-H on the same face as Fe) can be converted by acidic isomerizations with suprafacial H migrations into new structures, as shown. The other isomer (**114**) is merely racemized by equilibration with its enantiomer. Complex **114** also undergoes, with trityl cation, stereospecific hydride loss on the less hindered face opposite to the $Fe(CO)_3$, to form the symmetrical, racemized **117**. The set of formulae exemplifies some of a complex of mechanistically explicable interrelated transformations of terpenoids that we have examined. The correlated absolute configurations of the components agree with those of monoterpenes of known absolute configurations that I first determined in 1950[58] for quite other reasons.

Other Metals. We also examined some useful reactions of $Cr(CO)_6$ with typical Birch-reduction methoxydienes.[142] This investigation led to a new indirect process for efficient removal of OMe from an aromatic ring: initially by Birch reduction, then by reaction with $Cr(CO)_6$, and finally removal of Cr by autoxidation of the resulting (desoxy) $Cr(CO)_3$–aromatic complex. One of the synthetic examples was the efficient overall conversion of estrone methyl ether into 3-deoxyestrone.

Some Synthetic Applications: Shikimic Acid and Gabaculine

I realized rather late in my career that the "obvious" general applications of my synthetic methods were less obvious to others than to me, in the absence of specific applications. I provide here two examples of the utility of lateral control for rather simple but biochemically important structures hitherto obtainable only with difficulty.

The starting material in both cases, obtained by my favorite reaction, was 1,4-dihydrobenzoic ester (**118**). It was converted into a mixture of double-bond isomers of carbomethoxycyclohexadiene–$Fe(CO)_3$ derivatives, which is readily rendered uniform as **108** by acid treatment.[143] This sequence illustrates several aspects of lateral control of the organic structure and reactivity. Base-catalyzed isomerization of uncomplexed **118** itself, by contrast, only goes as far as the 2-carbomethoxycyclohexadiene. Moreover, with $^2H(+)$ as the reagent, the products after resolution [as the carboxylic salt of (+)- or (–)-phenylethylamine] are **108** ($R = {}^2H$) and **109** ($R = {}^2H$). Each of these products has a resolved asymmetric center on the ring carbon in opposite senses, due to ^2H versus H. To ascertain the absolute configurations of **108** and **109**, the CO_2Me in the complexes was converted after removal of Fe into organic products of known absolute configuration. Once the absolute configurations of the overall complexes had become known, that of a new center stereospecifically formed due to H versus ^2H was directly deducible. This structure is as shown.[144]

Removal of the Fe with Me_3NO occurs readily in the series without interfering with the final organic products. With **108**, this leaves an optically active diene (**117**) of known absolute configuration and full resolution in which asymmetry is solely due to ^2H versus H. I know of no other method, apart from an enzymic one, that can in principle achieve this so easily and completely. This is one reason why I call the general approach inorganic–enzyme chemistry.

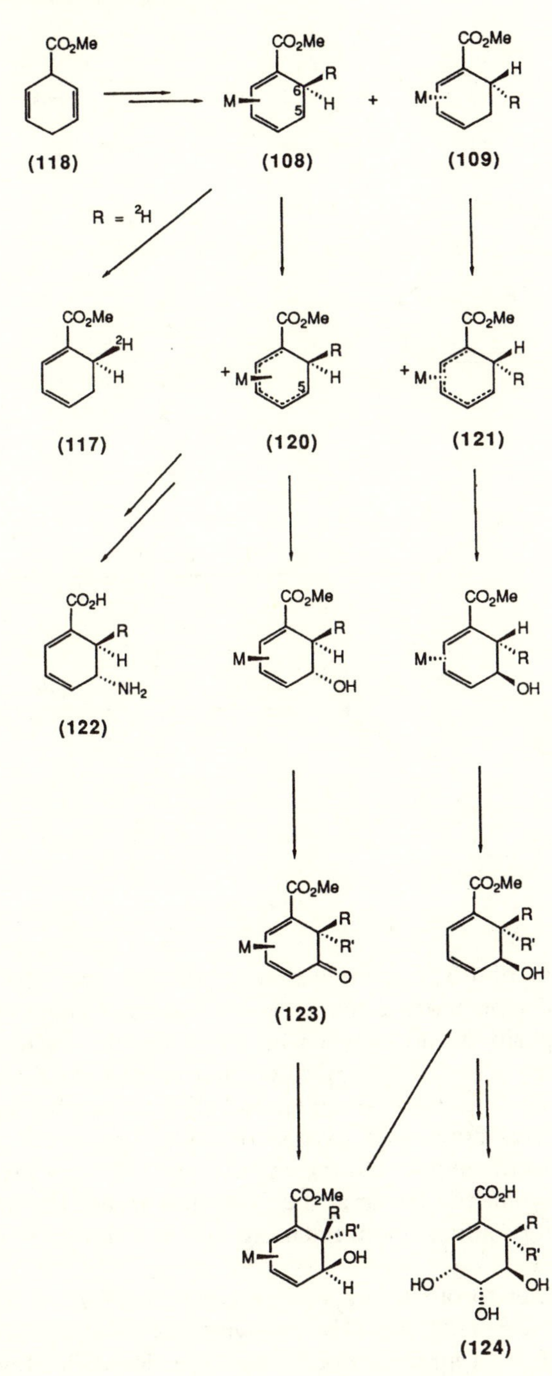
CO2Me
(118)
(108)
(109)
R = 2H
(117)
(120)
(121)
CO2H
NH2
(122)
OH
(123)
R'
O
HO
(124)
M = Fe(CO)3

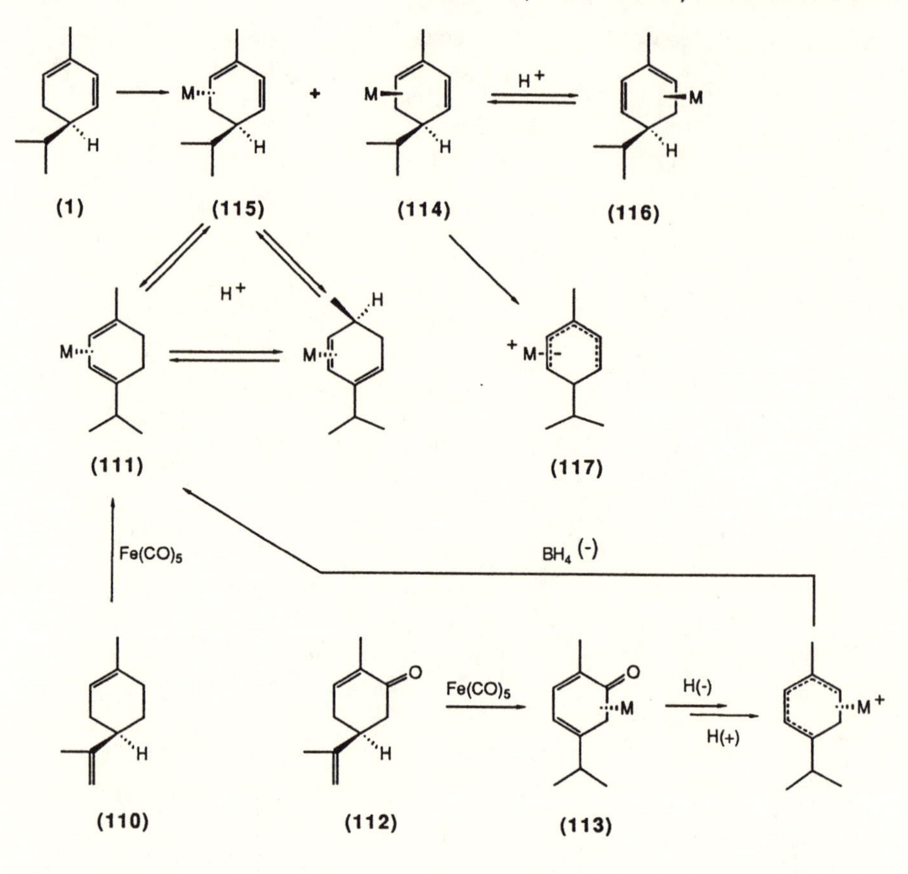

Reaction of diene complexes of type **108** and **109** with trityl fluoroborate demonstrates other aspects of lateral control of reactivity: the selective position and stereochemistry of hydride removal, and the stabilization of the resulting optically active mesomeric carbenium ions **120** and **121**. These cations can be reacted with nucleophiles regiospecifically at the 5-terminus with total steric control of attack on the ring face opposite to the metal. On the chiral complexes such processes generate a new asymmetric center at the 5-position, in full resolution and of known absolute configuration, retaining that due to deuterium when R = ^{2}H.

The routes to our two synthetic targets diverge at this reaction. With gabaculine[145] (**122**, R = H), the synthesis involves indirect introduction of NH_2 by appropriate protected N-nucleophilic attack at the 5-position of the correct enantiomer (**120**, R = H) and removal of protection. The success of the method not only makes this enzyme-inhibitor readily available (hitherto about $3000 per gram), but the known configuration of the precursor permits assignation to the natural material of its hitherto unknown absolute configuration. Another useful aspect of lateral synthetic control is demonstrated here because gabaculine itself is very unstable to alkali. The CO_2Me can be hydrolyzed to CO_2H before, but not after, the metal is removed. Complexation nullifies the electronic properties of the diene system responsible for the alkali instability. Enantiospecifically labeled compounds such as **122** (R = ^2H) can also be prepared in this manner for examination of the enzymic course of aromatization, "enzyme-suicide" reactions.

We similarly synthesized enantiospecifically (including with enantiospecific ^2H-labeling[146]) the fundamental biosynthetic precursor of one major class of natural aromatic substances, shikimic acid (**124**). The sequences proceed from the same cations (**120** and **121**), as shown. The method could obviously be used to make some substituted analogs for biochemical examination, starting from substituted benzoic acids. With ideas from my earlier biochemical work (Research Set 3), I had been hoping to use the resulting compounds in the partial biosynthetic formation of altered antibiotics. Alternatively, I hoped that they themselves might be antibiotic or useful enzyme-inhibitors in areas of plant biochemistry (such as *para*-aminobenzoic acid (PABA) formation).

An important feature of the lateral-control approach is that, although a resolution step is necessary, either the (+)- or the (–)-enantiomer of shikimic acid can be made from either of the resolved precursors **120** and **121** by a slightly different reaction sequence, as shown. The natural isomer **124** is directly available from **121**. Enantiomer **120**, which directs the hydroxyl at the 5-position into the "wrong" absolute configuration for natural shikimic acid, can have that center inverted to the "correct" configuration by extra oxidation–reduction. This transformation makes use of the stereospecificity of reduction of a 5-carbonyl in **123** with hydride attack opposite to the $Fe(CO)_3$. In the absence of ^2H labeling, both enantiomers **108** and **109** are thus convertible into the same final enantiomer (either natural or synthetic) of shikimic acid, according to the pathway chosen. In its presence (**108** and **109**, R = ^2H), these alternative procedures lead at choice to natural or to its enantiomeric shikimic acid, enantiospecifically labeled in the 6-position in opposite asymmetric senses for both series.

Such an approach makes a number of compounds available from the same starting materials by simple modification of sequences. To increase the resemblance to enzyme chemistry, it is necessary to find out how to make chiral complexes directly without a resolution step. We have been partly successful (about 40% ee) in transfer of $Fe(CO)_3$ from a

From left, Franz Sondheimer, Michael Dewar, and I at the 1978 Robinson Symposium in London.

complex of an α,β-unsaturated ketone, but this is one of the incompletely solved problems. Had I the opportunity, I would now like to explore another dimension of organometallic chemistry, aimed at the synthetic organic chemistry of transition metal clusters.

Random Conversations with the Editor

History and the Future

Seven years of discussions with the Editor, responding to his pointed questions and remarks, has encouraged me to address some important social and human activities of science, and of scientists as investigators and as people. My observations, as a participant and an interested broad assessor, may be relevant to interpretations of how scientific investigation "works" as a human activity and to its future progress. I have long employed a "conventionally uncommitted" historical viewpoint arising from my interest in the evolution of scientific ideas. I have traced this process in a major, essentially complete, collection of original chemical sources dating back to 1763 that I have assembled over a period of 50 years. My interpretations are based on my own experience. I do not use solely the usual secondary historical sources, although I also possess and digest many of these documents.

Some of the questions that follow are exactly as posed to me by the Editor. Others are really my questions, inspired by conversations with the Editor and advanced, based on my own fascinations and interests.

Editor: *Your viewpoint is not likely to be uncommitted; please explain what you mean by this term.*

My genuine detachment has two bases: scientific and personal. Scientifically, I have not been narrowly wedded to a specific field. I am habitually willing (indeed eager) to change, although in many instances this has involved abandonment of a fruitful field and worked to my disadvantage. My personal detachment must be largely innate; as a player in a match I always mentally observed the activity from the sidelines. My dedication to facilitating novel science is such that I always felt a strong sense of participation in the work of others whom I had helped to appoint and to present with organizational opportunities, as in the Research School of Chemistry. In that sense I was committed, but my concern was for science and for other people. I had a strong sense of obligation to my students who were writing theses, and their topics were largely chosen for their benefit as well as mine.

Editor: *The description of your career seems to infer that your choices of strategies were always rational, and indeed that the major outcomes were foreseen and virtually inevitable. To what extent is that impression an artifact of memory or presentation? How did you make your strategic choices? Also, in some of your account you come across as rather pretentious, which does not seem to fit your personality.*

I do not believe that my exposition is an artifact of memory or a selective discussion, as the published literature of the time can prove. As for being pretentious (that is, making claims), if that is the result of ideas that I alone had at the time, I plead guilty. I have not, I believe, set up incorrect myths about what I did and how I did it. Many of my practical circumstances were very hard. I had no margins for error in employing resources, so I had to choose very carefully. I suppose one reason for the apparently clear-cut nature and success of my decisions is that I do not have the space here to expound on all of my cogitations. But those ideas that I did adopt proved highly successful.

Why some people can dream rationally, and others not, is a profound question related to the personal basis of creativity. I do not believe that there is any formula except, before choosing a topic, to explore imaginatively as many paths as possible, to apply the hardest criteria, and to seek above all to prove each attractive proposition wrong, not right. If a proposal cannot be proved wrong experimentally, there is a chance it might be right.

Many novel ideas become apparent in dreams, as I discuss later. Kekulé said, *"Lernen wir Träumen—da finden wir die Wahrheit"* (Let us learn to dream; there we find the truth). These dreams can often be traced to subconscious ideas, absorbed but not consciously recognized. I have even wondered, for example, whether my polyketide ideas may have unconsciously expanded Collie's ideas that I had read in passing many years previously, with no interest at the time. But I have no conscious recollection of ever having heard of him or of his work when I evolved my ideas.

Many scientists fall in love with their own ideas and are inflexible about dropping them, whatever the evidence. They also allow the currently accepted paradigm of a period to color their whole approach. I was never like that; I was an eternal heretic. A paradigm (like phlogiston) usually has to die out with the people who believe in it (like Priestley). For some egotistical reason that must be innate, I never cared about any authority except my inner one.

Editor: *What factors provide the visions and the motives to pursue personally the strenuous labors of creative science? What are the necessary methods?*

Each individual has a different mix of motivations and capabilities for creative endeavors at the highest levels, but high emotion is a driving factor. To be in love with a topic summarizes the attitude for me. The basic personal reward is an appreciation of beauty within the complexity of the universe; it can take the form of an emotional search for order or an appreciation of novel creative disorder. Research involves dreams, but also the personal persistence to accept the hard practical grind, the search for exactitude painfully acquired, the boredom of information collection, the capability to continue in the face of disappointment, and the obsession to set aside other important matters. Sometimes it results in an unhappy personal situation, but it represents an unavoidable compulsion.

Sanity and progress require that one be able to accept the unavoidable defeat of a cherished idea (and to start again). I always set out to test my favorite ideas to destruction, but seldom succeeded in *published* work (nor did anybody else succeed). This attitude contrasts with the frequent (human) attempts of scientists to support their ideas. (I always hoped, of course, that in the end I would do so!)

Some people derive satisfaction from accumulating data, whereas others are content to dream and leave experiments to colleagues. Still others flit from flower to flower rather than learning more and more about one situation. The difference in approach is a matter of temperament, and we all must understand our own strengths. All workers ultimately contribute to the matrix of facts, ideas, understandings, techniques, and visions that we know as science. Financial rewards and moral imperatives on behalf of humanity have undeniable motivational value. However, I still believe that, whatever the exact ingredients in any particular case, the major impetus in scientific research is emotional. An additional factor these days is the interdisciplinary approach to problems. It flows from the desire and opportunity to collaborate with colleagues who make use of quite different techniques. I tended to make my choices quite independently, apart from collaborators who shared a training similar to mine. I consulted with them at my own choice.

Methods flow from ideas translated into technological action. They require a practical, inventive attitude (engineering, if you like) based partly on physical and biological methods and understandings. The objective is to translate broad novel ideas into practical actions, tests, and new detailed discoveries related to the whole physical world. The need to use physical instruments in chemistry demonstrates the unity of science.

Most scientists decide realistically, after they have had the opportunity to test themselves in practice, where their strengths and weaknesses lie. All types of workers are needed as contributors, but the highest level of research is an elitist pursuit. Probably 95% of breakthrough ideas come from 5% of scientists, but the work of the others is a practical necessity. Once, before computers, I had a staff member who spent 4 days per week entering filing cards of new literature. The task left him one working day per week to do something of his own. Nobel prizes are based on the illusion that a few individuals are more important than the cumulative advance of scientific understanding and techniques.

Editor: *You repeatedly use the word "creativity". Can you explain what it means to you?*

The word does not necessarily imply highly abstract intelligence, although that helps. (But how can we realistically measure abstract intelligence?) Rather, creativity involves the ability to make connections between facts and ideas that do not at first sight seem connected, and to use such connections to generate broad new insights. In short, creativity is the ability to see the obvious over the long term, and not to be restrained by short-term conventional wisdom.

Can creativity be cultivated, or is it innate? I do not know. I have a feeling that it is innate, but I am sure that it can be demolished by compression into conventional educational and social boxes. It is a paradox that the foundations of conventional scientific learning, and therefore of scientific creativity, must be laboriously acquired; thus both perspiration and inspiration are needed. One valid approach is that of the self-reliant heretic, questioning everything within an accepted order; the "Doubting Thomas" who needs to put a hand into the wound!

The creative pursuit of the knowledge of the universe as truth and beauty, contrasted with the often-egotistical methods used by its more mundane pursuers, is an index of a dichotomy that could be expected in an activity for which evolution presumably did not design human nature. But egotism provides the working potential for progress, and creativity thrives on a personal feeling of superiority (superego?) that often results in paradoxical obligations to behave unselfishly.

Editor: *What are the springs of successful ideas?*

In my view, ideas do not spring from a vacuum. Workers must have initially soaked themselves in broad areas of fact and acquired techniques. But for breakthrough novel ideas, a boldness in applying the matrix of ideas generated in one technical area in new ways to other areas is often critical. The farther the apparent distance such fields are apart, usually the greater is the resulting novelty, and the less likelihood exists that the first intellectual step will be taken. A critical mental approach to generating novelty is independence and confidence in thinking, untied by the popular and the conventional. Unfortunately, the apprenticeship method of research training, although good for developing techniques, tends to inhibit such lateral thinking. The creative researcher must be a strong character, or have an exceptional initial supervisor with a lower ego or a higher superego than is characteristic of most scientists (or else one who leaves you alone). Many doctoral candidates complain of supervisory neglect. To me this dependence implies the lack of the requisite ability.

In trying to assess how to pursue a topic, I found it necessary not to assume that practical science really works the way the published stories are told in scientific papers, in the form of logical, abbreviated sequences. Official historians, who are usually not scientists, at least in the area they are discussing, usually go wrong at this point. From my personal knowledge, many written accounts contain *a posteriori* rationalizations of the pathways actually taken; they briefly and logically recount what should have happened, rather than what did happen. Some chemists who boost their egos and some editors who save space by demanding short, logical presentations have a lot to answer for, didactically and philosophically. Of course, limitations on length prohibit details that are then considered irrelevant, but when their significance is recognized the real stories should be told.

Editor: *How do ideas manifest themselves, in your experience?*

Generating new ideas is an intellectual–practical activity based within the individual; all I can tell is how it works for me. The first point is that I have never classified myself. I am an organic chemist because I am interested in organic molecules. However, I placed no bounds on appropriate research subjects, including an early paper on the colors of Jupiter possibly indicating free radicals and ammonia, a view that has since become rather popular. I used to read the technical scientific literature when I was young, but later had time for very little broad reading; I tried to stay ahead of it. The major point, in my view, is not to label yourself as an alkaloid, a steroid, a theoretical, or any other adjectival chemist.

One very important set of introductions to ideas came when I was forced to do what I thought of as scientific journalism: to write reviews on rather unfamiliar topics. In 1950 and 1951, for example, I was asked to review the progress of the whole of alicyclic and aromatic organic chemistry over a number of years (the topics had been neglected) for the Annual Reports of the Chemical Society. I included within all such surveys new ideas and correlations, including the first set of absolute configurations of monoterpenes (based on my idea of using these to define phytochemical intertransformations) and also some new natural-product structures in fields in which I had not worked. Thus I first became fully aware, through summarizing them, of the courses of Diels–Alder additions, which would be important for my later synthetic work. I was asked to participate in CIBA Foundation medical–chemical discussions in the early 1970s. My role as a chemist, I was told, was "to keep the medics straight". I was notably stimulated by the unfamiliar topics.

George Kenner (right) and I "testing sherry" for the CIBA Foundation, about 1970.

This responsibility forced me to look at many biologically related problems and new (to me) biosynthetic ideas. This effort led me to define the importance of N-oxidation processes (e.g., in the biosynthesis of penicillin), some of which are discussed in Research Set 3. A number of other invited reviews for books and conferences, for example on lignans, were equally important to my general novel thinking and new scientific results. My mind does not work well in a vacuum, and I need some practical stimulation, particularly if I am to consider a new area seriously. Also, deadlines overcome my laziness and tendency to boredom.

A common factor in all of my major research areas is that they normally started with one requirement, or one observation, which I followed up, and which expanded, usually in unexpected ways, sometimes almost explosively, from a definable center.

Such starting points (or "creation centers", as I privately named them) could become established in a variety of ways. The recognition of one always led me to do something in the laboratory, an approach that I found to cause a radiation of further creative ideas. For example, Robinson provided me with one center: the necessity for steroid total synthesis from a specific aromatic precursor (**18**, Research Set 2). This requirement radiated as I have described into some major synthetic methods and ideas. One method in particular, metal–ammonia reductions, itself became a new and very major creation center. From its radiations came other new centers, based on the one hand on recognition of new types of synthetic approaches made possible by the availability by reduction of classes of materials like methoxycyclohexadienes. It also led to the formulation and solution of theoretical problems related to the revealed need to rationalize the addition of electrons to molecules, and to the fundamental mechanistic (and synthetic) aspects of the reactions of the resulting mesomeric radical anions and anions. Without a practical background I would never have sought the theoretical insights. Other people clearly have quite different approaches, desiring the solution of purely practical or of purely theoretical problems for their own sakes. I found that it is often helpful to be doing something practical, no matter what, to permit visions to arise, rather than to await the advent of a grand theoretical overall plan. I cultivate "feelings". My resulting theoretical developments led in turn to new practical centers, such as the first deliberate deconjugations of α,β-unsaturated ketones and acids for synthetic purposes. These deconjugations followed from the distinction I had necessarily made, from my reduction work, between products of kinetic and thermodynamic control. And so my theoretical and practical webs expanded and continue to do so, even into organometallic chemistry, which is discussed in Research Set 4.

Editor: *But how do you define individual problems to be attacked?*

A specific research problem could be defined the conventional way, by sitting down with pieces of paper. Another way, which tended to lead either to novel strategic ideas or to nothing at all, had several necessary precursors. I had to have soaked myself in general information (for example, by a review operation as noted for the Diels–Alder reactions) with a broad interest rather than any specifically defined purpose, although a variety of such purposes needed to be hovering in the background. All of this background I seemed to digest subconsciously. Then, at a particular moment, I had to arrive at a "gut feeling" of excitement that something novel was ripe. I said to myself, "Let me think about what to do in this area." I let my mind float freely in a sort of undirected whirl, into which I could feel myself almost physically grasping. This feeling could happen anywhere, sometimes during discussion with a collaborator, sometimes in a boring committee meeting. There was nothing consciously rational about it. After a minute or two, some idea either emerged or it did not, and I could try again the next time I became irrationally excited. It is significant that the reasons for my resulting thinking could instantly be rationalized after they had become evident. I think I know how Kekulé felt in generating his own pregnant creation center from the Theory of Types.

Editor: *How do you organize to convert dreams into practical realities? What is the role of teamwork in science?*

I was never a very good teamworker, except as director of the team. Similarly, I was an effective member of a committee only when I chaired it. When necessary I could adapt, particularly when other team members represented quite different areas of expertise. In that case we could all contribute questions and answers reflecting varied insights and practical approaches. Among my students and close collaborators, I often preferred those who could make ideas work, rather than those who generate startling new ones. (It is a different kind of creativity, not that I rejected ideas.) The combined efforts of complementary types of researchers are needed for practical progress in modern times. The overall desirable attributes, within individuals and in various mixtures among individuals, are creativity; interpretability; skilled experimental, organizational, and entrepreneurial abilities; and a tendency to be emotionally moved by combinations of visions of understanding and of human application. The resulting total activities unite discovery (uncovering and understanding what is there) with invention and development (making conscious use of discovery for human purposes, including more discoveries).

Appropriate collaborators sought for a given purpose will supplement or reinforce your own characteristics, but they must be allowed to develop individually. I have been fortunate in mine. It would be invidious to be selective, but I must mention a few. My first senior collaborator Herchel Smith supplemented in a most effective manner my then-unbusinesslike attitude to details of organization and to writing papers. Rod Rickards was (and is) highly creative, both in developing ideas and in working them out. G. S. R. Subba Rao was very effective in a typically indefatigable Indian fashion in reducing some of my airy general suggestions into specific practice. More recently, at the ANU, I have benefited from the experimental work and ideas of a number of brilliant collaborators, among them A. J. Pearson, G. R. Stephenson, P. Westerman, Acharan Narula, and Larry Kelly. I have tried to contribute what I can strategically to a common cause, and to help bring out from others what they have to offer, which is usually a great deal.

I have referred elsewhere to some of the problems encountered with organizing equipment (e.g., in mass spectrometry). A combination of scientific and entrepreneurial activity is needed for most scientific achievements, and initially I lacked it. I acquired equipment not because it was a good idea, but because I needed it for a defined purpose. In one university (Keele), where I had some formal research supervision duties (Manchester), a supervisor caused an unfortunate Ph.D. student to spend 2 years building a standard peptide-analysis machine (not commercially available) before he realized that he had forgotten to provide

Ganugapati S. R. Subba Rao, Jessie, me, and Lakshmi Subba Rao at a temple in Halibid, India.

a problem. This I had to do from mercy, quite outside my own technical area; I lost the scientific results but learned a major organizational lesson.

Editor: *What were the basic underlying motives that drove you all these years?*

I think I have answered this question in parts of many other answers.

But, Editor, by now you must know a lot about the psychology of scientists. Can *you* tell me why I was so irrational? My dedication resembled love: Why do people fall in love? After experiencing scientific research, I could not imagine doing anything else. The ego reinforcement that accompanies success undoubtedly helped to prolong the attachment. If I had been unsuccessful, I suppose I would have accepted mundane routine scientific tasks in order to earn a living. As it was, I just managed to scrape by financially. I also like teaching, if not examining. The feeling of being able to spark the enthusiasm of young people, especially undergraduates, has been a continuing stimulus for me. I feel the same with my grandchildren. I repeat, scientists are egotists, but this characteristic takes many forms. In my case it included a sense of *noblesse oblige*; because I happened to have the ability, I accepted the implied obligation. This "puritanical" viewpoint is now unfashionable.

My main motive was to satisfy my curiosity about the world. Because I could not investigate *everything*, I chose to delve into one field that appealed to me. As a boy I was stimulated by colors, smells, beautiful crystals, beautiful laboratory glassware, fascinating changes and precipitations, and a sense of penetrating below surface appearances. I wrote a recent update by the Drs. B.[147] of the highly influential early 19th Century *Conversations on Chemistry* by Jane Marcet. Her original volume converted Faraday to science and caused Mary Shelley to react by writing *Frankenstein*. I was struck by the interactions of Mrs. B. (who knows everything) with the eager teenager Caroline (who likes the bangs and smells and burns holes in her clothes) and the serious Emily (who wants to know what it is all for and how it works). I saw myself as a combination of those two students, but I lacked a personal Mrs. B. except in books.

Editor: *You speak of right and wrong conclusions. Are these not relative terms that can be assigned only within environmental contexts, not absolute terms? How do you know?*

This is a deeply philosophical question, one that even religious leaders and philosophers cannot answer. I once talked to the Sri Sai Baba (allegedly the reincarnation of the 18th Century prophet Sai Baba, but also of Krishna, Christ, Mahomet, who should have known if anybody), "but ever by the same door came out wherein I went", as said Omar the Wise. I have tried to set up scientific criteria for ethics, but always end by asking the meanings of words. I take a pragmatic view, and I agree that in one sense terms are relative to a time and a set of knowledge. Newton was "right" in his time, as Einstein was in his.

But I believe that some things in the universe are *there*, even if we cannot comprehend them and perhaps never will. (Many philosophers would disagree, but I am a scientist.) For example, I assume that the molecular structure of strychnine is *there* and that our knowledge of it is true. It can only be falsified by a redefinition of structure, although it can be refined in terms of increasingly exact atomic positions and electron distributions. To "disprove" it at this stage would imply disproof of the million or more organic structures known. I do not believe in such improbabilities. My criterion of truth is that it fits all known facts (but what are facts?) and can be used to make verifiable predictions. Wrong conclusions are to be denied by experiment, even by one valid result to the contrary. Initially, in postulating possible truth, I have looked at whether a conclusion is probable. It "looks right" was a favorite phrase of Sir Robert Robinson about a new alkaloid structure. This judgment was based on wide knowledge and an ability subconsciously to select and to incorporate important factors. He was almost always right.

My views, of course, are most easily applied at the molecular level of organization and above. Because much of physics is so "fuzzy" to human understanding, other rules may seem to apply.

Editor: *A favorite statement by scientists in support of uncommitted research is that most breakthrough advances result from chance. What is your experience?*

In one sense that is true, in another not. For "chance" to occur (*chance* in Pasteur's French means luck) you need to have the right attitude of mind to perceive the significance of unexpected events, and that attitude does not come from luck. Many of my breakthrough ideas—like the Birch reduction—have arisen from basic strategic research (i.e., an effort to satisfy long-term human needs) rather than from totally uncommitted research (contributions to the discipline). Nevertheless, they have also provided major clues to fundamental advances in science per se through the recognition of *chance*.

Ernst B. Chain once told me that his unique and critical contribution to the work on penicillin was due to an accident. In the intervals between looking for an apartment that could hold two grand pianos, he came across Alexander Fleming's paper on penicillin. He had intended, at Howard E. Florey's suggestion, to look up the paper on lysozyme. He then persuaded Florey that it was a more practical approach to an antibiotic. At that time, Chain was fortunately unaware of Raistrick's practical failure to isolate penicillin. The knowledge might have discouraged his enthusiasm, but I doubt it. He worked on penicillin for several years before he realized that it contained a sulfur atom. This revelation followed a query from an analyst about the nature of a peculiar residue in the combustion boat: in formula, 1S = 2O. His crucial intervention was "accidental", a combined result of racial politics (as an exile from Berlin, he was willing to work at any task), Florey's ability to judge people and projects, and the lax indexing of the period; but that is how science often works. I customarily read not only the papers I needed, but any others that were of interest.

Editor: *What is the value of spending increasing amounts of time, money, and human resources in defining small, particular aspects of science more and more precisely?*

Many governments would like an answer to this question; if I knew the answer I could make a lot of money! I faced the question on a large scale when I chaired the inquiry into CSIRO, the principal Australian government research organization, and when I set up funding and projects for the Australian Marine Sciences and Technologies Research Committee. In a 1976 attempt to clarify the situation I invented new classifications to replace the imprecise terms pure and applied: fundamental research (either strategic or uncommitted) and tactical problem-solving. My classifications form a matrix, rather than a hierarchy of prestige. There is frequently no technical difference between strategic research (which in my case led to the Birch reduction and the 19-norsteroids) and uncommitted research. The difference lies in the reasons for project selection (to form the novel basis of a problem solution or to contribute to a discipline, although they usually do both) and the fact that strategic research has to be pushed to a conclusion. One reason for the attraction of uncommitted research is the possibility of following up on unexpected new discoveries, which often leads to easy publications.

I wish I could give a clear answer to the question. A partial response is, "In research you never know what will be important." Another explanation favored by some scientists resembles the argument for climbing Mount Everest: "It is there, and it is possible." Competitive allocation of resources is necessary in this expensive new world. My own organizational solution was to provide some (not necessarily generous) funding for uncommitted research. This uncommitted funding was to be granted on the basis of assessment of the people involved as much as the proposals. This approach permits genius and accident to open new vistas. I supported success. I did not, for example, spend money on equipment because it "seemed like a good idea" or "everybody has it", but because I could clearly foresee types of specific need.

A typical political approach, exemplified most strongly in the old Communist countries, is to make some sort of half-informed decision on the basis of what is desirable, neglecting what is actually possible. Large sums of money are wasted in this way, for example on pedestrian aspects of cancer research. I recall, while serving on a U.K. medical committee, rejecting a request for funds to visit the United States and for a study of the statistics of cancer of the left breast in Oldham. A combination of scientific and political wisdom can open new possibilities, but eventually the effort must be accountable. Unfortunately, most decision makers are "famous for being famous" and tend to go along with political power brokers.

I accept the need for a broad definition of areas, but then let the people with ideas move ahead! Although sharply limited research topics may eventually fit into a larger picture, I have never favored scientific "stamp collecting". I always want an indication of what the whole picture is likely to be.

Editor: *What are those "certain principles that pertain to the profitable conduct of research and to successful interaction with co-workers"?*

Most of this topic has been covered, both in other answers and in the main part of this book. Revolutionary ideas are the products of a few people, who must be sought out and encouraged. But these earth-shaking ideas are ultimately based on ordinary science conducted by the mass of scientists, laboring away patiently and making sense of their findings. In that respect, research is partly "top down" (notably in the making of strategic choices on the basis of experience) and partly "bottom up" (in making use of the new ideas and novel techniques contributed by younger people). The organizational problem is how to marry the two approaches, so that the top people will not dominate with unwise choices and the masses will have an opportunity to reach the top if they are qualified. The extreme solutions are to treat the professor or research director as a god, or to settle for unexciting consensus opinions arrived at by the democratic masses.

I tried to solve this dilemma by maintaining direct contacts at all levels. I seldom saw students in my office, except at their request, but frequently sat on a stool in their laboratories. I was prepared, on their behalf, to stand up to bullying administrators with axes to grind. I treated the technical and support staff in the same way as the academic staff. This policy sometimes led to unfounded protests that the technical and support staff were making academic decisions. During 1979–1987 I made six visits to the Peoples Republic of China, some of them with my laboratory manager John Harper. With a focus on technical and laboratory management, we clearly defined the functions and responsibilities of the research staff at all "levels". Judging from the Chinese enthusiasm, they will be teaching us soon.

My objective with the Research School of Chemistry was to justify our flexible funding by demonstrating the success of all the workers in the school. In terms of basic approach, as foundation Dean I set up a list of types of research appropriate to the Australian general environment:

- inorganic and organometallic chemistry, because of the importance of minerals and gas in Australia
- organic chemistry, notably related to natural products and primary industry
- insect, marine, and fungal chemistry of world importance, but presenting specific Australian opportunity
- insufficiently studied aspects in Australia like basic theoretical chemistry
- topics like organic synthesis, which could be turned toward many aims

All of this research was to be carried out at the highest possible achievement levels, as contributions to the world of science. Achievement of this goal involved setting up instrumental techniques at research levels. The work later developed appropriately into subjects like solid-state chemistry and new materials. The School is so flexible that it can rapidly react to new opportunities and obligations. That situation reflects my vision of the profitable conduct of research, which automatically leads to useful practical outcomes, not the least of these being the self-training of excellent people.

Editor: *A considerable problem for junior collaborators seems to be how to free themselves from their supervisors and acquire independent reputations. How did you handle that problem?*

Initially I gave my collaborators as much choice as possible to work on what really interested them, with many topics presented. The chief bureaucratic problem for my students was to find an initial title to submit for their theses. Every creative scientist knows that if you start for New York you may finish in San Francisco; that is the way of research. But administrators demand initial titles for theses, and these must appear on the final version for the examiners. As a result of conflict with bureaucrats, I became an expert at framing titles that permitted the widest possible interpretations. It was an amusing game for me, but the process often freed students for real research.

A more serious personal question concerns postdoctoral collaborators. They are usually appointed within a particular field to assist a particular supervisor. But, because of the stringency of their selection, they frequently show outstanding creative qualities of their own. Should the supervisor claim the credit for publications of these collaborators simply because his or her name is on the papers?

This dilemma is not straightforward. If a supervisor suggests the field and the strategic approaches, tactical contributions can be taken for granted. But what if the postdoctoral collaborator comes up with an entirely new and creative approach that leads to significant results? My own puritanical and moral attitude often worked to my own disadvantage. I generously gave such people the freedom to publish individually, as soon as it became clear to me that they were working on their own novel ideas, rather than mine. Why those ideas had been generated was irrelevant. If I had been an old-fashioned research director, my name would have been on many more important papers. I never favored the big-battalion approach to demolishing problems; my weapon of choice was scalpels, not bludgeons. In contrast, I knew one distinguished research director whose criteria included whether he had organized the grant, let alone the science.

Real-life research supervision leads to another, wider, question. If, as a supervisor, you "point the nose" of collaborators in a certain direction, you must expect that later they may choose to continue independently in the field. I always accepted that fact. Still, I occasionally became rather annoyed when a former collaborator (who must have known what I was projecting) took credit in print for what I had in fact already done. The moral, I guess, is either to keep your ideas to yourself (which I abhor) or to be systematic about publication (which many creative scientists are not). My goals involve promotion of scientific progress, which excites me even if I am not credited with it.

On publications I listed the author names alphabetically, as a sign of equality. (Of course, my name begins with B!) The alternative would be generating random order (which might work with a large number of related publications) or trying to assess the relative contribution of each author. Alphabetical order has its drawbacks; my old organic chemistry mentor Frank Lions, who largely provided the initial idea of making polydentate chelate complexes and all of their theoretical implications and who initially made the organic pieces, receives little credit. All of the papers reporting this work listed Frank Dwyer first (he was an equal collaborator). Besides, it was considered inorganic chemistry, and Lyons was an organic chemist.

I often included research assistants and technicians in the author list, unlike old Professor Charles Fawsitt in Sydney. He lost one eye in an experiment and never risked performing another. Instead he thanked his technician in footnotes for all of the experimental work in the paper. The students called his technician "Weel", from Fawsitt's habit of saying, "We'll do this, that, or the other", and then wheeling in "Weel" to do it.

Editor: *Please explain the role of "supervisory control" in the conduct of science. As Robinson's student and while in his laboratory, did you appropriately follow his guidelines in the research activities you undertook?*

I have already referred to aspects of this. Robinson was an ideal supervisor for me; he left me alone. For others he was not so ideal because they required more tactical guidance than he was able to give. He suggested (with E. Stenhagen) my boring Ph.D. topic: long-chain fatty acids containing a quaternary carbon atom, imagined incorrectly to relate to tuberculostearic acid. Robinson left me alone to devise methods for making them. I came up with a number of general methods for producing quaternary carbon atoms, which were later of use for steroid synthesis (angular Me groups). I found these challenges more interesting than the alleged problem. Because I infrequently bothered Robinson, he hardly knew what I was doing. He seemed uninterested except when I brought him a written paper. I later regretted my lack of importunity because in the few discussions I had with him, usually on a problem referred to him by somebody else like H. Raistrick or J. L. Simonsen, he impressed me with his rapid creative reactions.

Robinson was unique in his time and was apparently willing to accept any responsibilities thrust upon him (he served on 37 committees during the war). Unlike Alex Todd's action later in similar situations, Robinson refused to appoint assistant directors of research. One initiative in that direction (not by me) was angrily rejected. The Robinson approach to supervision, I later learned, was common in the arts, but not in science.

I have discussed elsewhere the origins of the 19-norsteroid work. Robinson suggested the steroid analog problem in the form of cortical hormones and financially supported me (350 pounds per annum). From my point of view, his critical contribution was the suggestion of his ring-A aromatic simplified steroid precursor as a major starting material for synthesis, although he had no idea of how to use it. This challenge was an obsession for me for many years, and was finally brought to fruition in hormone total syntheses. His greatest influence on me came through my observation of the creative way he thought (which I tried to follow) and of his administration (I tried to do the opposite).

Editor: *What is the role of "fun" in chemistry?*

That depends on the teacher and the system. I have frequently heard research chemists say they are having fun, or that a particular result is amusing, or that they are playing around. I have said so myself in summing up my personal reaction to what I am doing. What is implied? The word "amuse" originally implied an open-mouthed stare, presumably of amazement, usually associated with a smile. "Fun" comes from fond, folly, playing the fool. A sense of humor implies also a sense of proportion about your own importance in deriving a result. Science is not an inhuman search for knowledge, but something that fulfills emotional needs of the investigators, one of which is fun, however you define it. But to have fun, you need to acquire (often painfully) the fundamentals.

Conscious humor is perpetrated on occasion, for example in producing names for new natural products or new concepts. This humor often seems to me rather ponderous, resembling the activities of the cartoon hippopotamus ballet, dancing as a chorus line in tutus. However, jokes can have a serious purpose beyond play and relief from tension. For example, my old friend Jack Edwards (Montreal) (under the transparent pseudonym of Aloysius S. Smith) published in *Chemistry and Industry* a parody of the uncontrolled trend toward all sorts of curious names for sterically directed bonds. It started as an apparently very serious proposal. About half-way through the paper, one started to smell a rat (such as in the reference to Eni Meni and Mini Mo, *J. Hotsitotsi Inst. Jpn.*) that became increasing malodorous. The inevitable final question after the amusement was if serious proposals in the area are equally ridiculous.

John Cornforth's unpublished chemical limericks are outstandingly witty, and good mnemonics. Nobody is likely to forget the meaning of chiral after what happened to a strategic portion of the anatomy of a stereochemist of Tyrol who slid down a spiral. I quoted one earlier because it refers to me. I am under an interdict for more because he alleges, reasonably, that they are "occasional" and should not be quoted out of their creative context. One that I think can be squeezed into the context here came from the 1960 IUPAC Symposium in Canberra. At the dinner Todd sent a note to Kappa Cornforth, saying, "For God's sake write me a limerick; I have to make a speech." The instantaneous result (on the back of the menu) was

If you are anxious for over-exposure,
to prepublication disclosure,

to good food and good drink,
without leisure to think,
try IUPAC Symposia.

I recall with amusement the meetings of the Syntex consultants board. Members at various times (along with George Rosenkranz, Alex Zaffaroni, Bert Bowers, John Fried, and others from the firm) were Howard Ringold, Charles Sih, E. J. Corey, Franz Sondheimer, Koji Nakanishi, Gilbert Stork, Carl Djerassi, and others. The meetings sometimes resembled gladiatorial contests with loud disputation but good humor, bets being freely exchanged about predictions, notably between Gilbert Stork and Carl Djerassi. The bets were even collected later, and I recall Carl framing a $10 bill that he, somewhat unusually, won from Gilbert in a particularly hotly argued case. This attitude kept everybody on their toes with interest in what could easily have become drowsy meetings. Between bouts, Gilbert used to catch up on recent issues of journals, to the freely expressed annoyance of Carl.

Scientists can have a sense of humor, related to a sense of proportion, in personal affairs. Robinson's pleasant side was evidenced by

"Birch Day" at Syntex Corporation in Palo Alto, California, in October 1987, marking the end of my many years as a consultant. Left to right: Gilbert Stork, George Rosenkranz, Carl Djerassi, Albert Bowers, Alex Cross, John Edwards, John Fried, and me.

John Edwards (right) and I at a Syntex Corporation lecture around 1975.

his music and chess. He left most social contacts to Gertrude, a very kind and sane person despite caring for a mongoloid son, a situation that was a great trial to Robert in particular. He liked pleasing people, when it occurred to him. I was in his office when the secretary rushed in and said, "Dr. *X* (a Canadian collaborator from some 25 years before) is here and would like to see you." Sir Robert responded, "Quick, get me the reprints for 19*XX*." When the lady was ushered in he said, "How nice to see you again. Was it not remarkable that the anthocyanin we worked on turned out to be a mannoside?" She went away happy, thinking that her remarkable result had been in his mind all that time. Robert chuckled happily over his little piece of gamesmanship.

Editor: *What are the influences of personal interactions on scientific training and attitudes?*

Research is, in theory, taught by the apprenticeship system. Apart from broad attitudes inculcated by observing Robinson and Todd in action, I learned technically from fellow students and later from collaborators. I never had any qualms about about consulting experts or revealing my ignorance of the fine points of "new" techniques like NMR or my terror of computers (both invented long after my years of training). In return, I tried to point my students in strategic directions identified through my experience; then I left them to make their own achievements and mistakes. One of my many dicta was, "If people are not allowed to make their own mistakes, they will never make anything worthwhile."

Research is a wide social activity, often involving what a German friend of my called *Friende* (friendly enemies). Conferences influenced me not so much through information imparted as through the excitement of new achievements that otherwise I might not have encountered. The people who succeeded were faces, not pieces of paper with informal discussions *à batons rompus*. They act as a stimulus to emulate (not imitate) achievements. Many university administrators and politicians do not seem to understand the importance of personal interaction in fostering creativity. (I recall Malcolm Fraser, our Prime Minister, talking of "scientific jaunts abroad". "Oh indeed, Prime Minister, I assume that does not apply to politicians?")

My fellow students in Sydney (1933–1938) were a brilliant lot not only in chemistry but in mathematics, physics, and biology. The graduates in the 2 years before World War II have probably not been equaled since. The atmosphere was stimulating and competitive.

Rita Harradence (Lady Cornforth) was in direct competition with me. She had a more distinguished academic record than either I or her future husband. I suspect that my "1851" in the sexist days of 1938 was awarded because I was male. Professor Earl, who was quite a politician, secured one for Rita (which she deserved) by putting her first in order before Cornforth so that they both received awards in 1939. Science is still unfair to women, perhaps because they tend to be less obsessively egotistical human beings than men are, although there are exceptions.

John Cornforth and Ron Nyholm were a year behind me. Perhaps this was fortunate because Cornforth's sheer brilliance (not just in science but in chess, sport, and literature) would probably have fatally discouraged me. When I first met him we were at different schools. He was a Greek and Latin scholar intending to read law, but with some interest in the spectacular side of science. Probably in 1932, we traveled on the Manly ferry and on a tram to Long Reef (North of Sydney), carrying tins of potassium chlorate and sulfur with detonators and fuses to collect plant fossils from the cliffs (Triassic, Narrabeen Series). I have shuddered since, notably when one of the laboratory assistants in Manchester, before my time, blew himself in halves. I once noted, in introducing Cornforth, that we had collaborated on exothermic chain reactions, but because of some uncertainties in the results they had not yet been published! About this time he blew the power lines in his street while constructing an electric furnace. I apparently absorbed heretical attitudes, but otherwise was little influenced directly by those around me.

Sir Hermann D. Black (left) at a ceremony conferring the honorary D.Sc. on Sir John Cornforth (right) and me at the University of Sydney.

Editor: *You have known many scientists. Are there observable connections between personality, life-style, general attitudes, and scientific creativity?*

What becomes scientifically obvious to an individual depends on national ethos and educational development; interactions with collaborators, with society, and with the international world of science; but largely on rather indefinable individual personality characteristics. The result is *style*, a difficult concept to examine other than instinctively or by example. In earlier days, I could often tell who had written a paper from both the work and the style of presentation, without even looking at the title page. The only confusion for me on such written papers arose from the influences on the principal collaborators of very powerful research directors like Woodward and Barton.

Life-styles seen early in a career may not be clearly cognate with scientific styles, because many people are initially constrained to behave as myths tell them they should. Temperaments eventually equate with achievements or else explode, sometimes with grave personal and career results. "Know thyself" in genuine honesty, warts and all, has been my aim.

It takes all sorts of temperaments to make the world of chemistry: discoverers, inventors, dreamers, practitioners, entrepreneurs, technicians, "stamp-collectors", and transmitters of knowledge. It is difficult to combine all such attributes in one person, because some imply mutually contradictory personal characteristics. Fortunately, many people seem at least to be able to alternate in several different roles (e.g., as discoverer–inventor–applier–teacher).

Robert B. Woodward

Bob Woodward's highly-thought-out style as a synthetic chemist was far more effective than Robinson's or mine. It affected the whole outlook of subsequent synthetic chemists. On the surface his life-style appeared similar to his work, very controlled, rational, and calculated. However, I suspect he was very emotional underneath, which is probably why he smoked and drank to excess.

Perhaps he had too much scientific influence, in that followers without his ability tried to do what he said he did, to think totally through a problem from beginning to end before even commencing experimental work. In print and lectures he had the gift of making a synthetic sequence look inevitable but not obvious. His brilliant lecture exposition appealed to the imagination, particularly through artistic illustrations (colored chalks, dusters, style). In my view, his published emphasis on prescience and rationality sometimes did some injustice to the high level of creativity in dealing with developing experimental situations that made his style inimitable.

Woodward did not seize on one highly attractive reaction and insist on building a sequence around it, as Robinson and I did. He demanded the experimental polishing of each reaction stage at any cost to himself or his collaborators, to permit attainment of an acceptable amount of final product even after 30 or more synthetic steps. He devised a microexperimental approach to accomplish maximum yields at each stage, using spectra to monitor products.

He was self-acknowledged as not a good experimental worker, and he judged his collaborators primarily by their experimental dedication (in order: Swiss, German, English, with Americans usually behind, excluding major personal exceptions). He once told me a story about personally trying to introduce the 19-angular Me into steroids via CH_2Cl_2, by reaction of phenolic estrone with chloroform and alkali (which theoretically might introduce either a 2- or 19-CH_2Cl_2 group). He had obtained a nonphenolic (19-substituted?) crystalline product when his wife telephoned and demanded that he go to look at a new apartment in Cambridge. By the time he returned the product had disintegrated, and he never succeeded in obtaining it again.

Often Bob's reactions were not as novel as they appeared, but they are exemplars of highly organized and consciously polished processes. Robinson once described the difference to me: "These Americans are like cats trying to get over a wall; they scramble and scramble until at last they reach the top. I walk up and down looking for a hole to get through." My reply: "Yes, Sir Robert, but supposing there is no hole, or you cannot find it?" But Robert had great faith in the existence of such holes; maybe in the very long run he was right.

One drawback to Bob's approach is that if a reaction did not yield the expected result it was discarded, or never published. In consequence, much information of genuine scientific importance could be lost. (I know; I read many of his unpublished research reports and theses in Cambridge, Massachusetts.) He told me, in response to my protest, that there was too much nonsense and triviality in the literature already. I asked, "Bob, how can you judge for the future?" But he believed that he could. It is perhaps relevant that he was an inveterate, and usually successful, poker player. He was prepared to lose, but seldom did.

I could not take myself or my work as seriously or obsessively as personalities and research were taken at Harvard. I recall being left by Bob to chair a seminar one evening, while he took the Cornforths to the airport. At about 11 p.m. the seminar had taken on some of the characteristics of certain types of religious meeting. People were contemplating their navels for long periods, presumably seeking inspiration. When nothing seemed to be forthcoming I closed it down, to great consternation; people refused to go away. Bob was not very happy when he eventually returned.

He could lecture for hours, to the embarrassment of later speakers at a conference. I recall that he delayed a reception by Prime Minister Menzies in Canberra in 1960. At an IUPAC Conference in Delhi, he was set down as the only lecturer on a Saturday morning for this reason. I asked him when he was leaving: "2 a.m. Sunday, immediately after I have finished my lecture." (This was a gross exaggeration, but indicated a sense of humor.)

As to personal obsessions: *"Was nicht in den Annalen steht"* (roughly, "Off the Record") in *Angewandte Chemie* once reported that Woodward had been seen wearing a red tie. (Everything he had was a shade of blue, including the ceiling of his office; Rorschach might know why.) When asked why he was wearing the red tie, he replied, *"Heute bin ich incognito."* ("Today I am incognito.")

Sir Derek Barton (seated, left), T. R. Seshadri (seated, right), and I at the IUPAC Conference in Delhi. There were too many gods in India, a statement not referring to religion. Seshadri was one.

Gilbert Stork

Of my more immediate contemporaries, Gilbert Stork is the one whose synthetic work genuinely dazzles me with its simplicity and inevitability. He devises, on fundamental bases, novel creative key reactions that are obvious for his synthetic purpose, and he has the charisma to persuade his collaborators to carry them through experimentally with great enthusiasm. He has a great sense of humor, not unrelated to a sense of proportion, and a personal deprecation that makes him one of the most widely liked and respected organic chemists. As an intensely creative and original person, he also has a very broad interest in and appreciation of art, including abstract and modern art.

Gilbert's life-style reflects this inner creative aspect. There are more car stories about him than about anybody I know. I suspect he may have embroidered them a little. In his own estimation he may be the second-worst car driver in the world. For instance, a wheel once rolled past him on the George Washington Bridge. "Somebody has lost a wheel ... I have." Question: "Did I have a flat the day before? Did I change the wheel?" His guardian angel works overtime, as this and many other episodes indicate. He introduced me to the subtleties of bribing the Mexican traffic police, among his other creative activities.

He has a great continuing interest in his students (many justifiably in high places) and in their careers. I wish I could imitate his style, but it is inimitable and very personal.

Franz Sondheimer

I was very fond of Franz Sondheimer, who demanded the best from life and from himself. I recall with great pleasure his magnificent scientific book collection through which I used to browse; his house at Las Brisas, Acapulco; converting traffic police to our favor in Mexico City, night clubs, and the Israel desert.

He embarrassed me once when he was lecturing in Manchester. The president of the student Chemical Society took us to dine at the "posh" Midland Hotel. Franz took one mouthful of each dish and ate no more, without explanation. I knew that the food did not measure up to his standards, and so did the students.

He was an individualist who at one time flew the only private plane in Israel. He always carried an ABC air guide so he could find unorthodox ways to fly between impossible places.

As a great gambler, Sondheimer was willing to back his luck and his judgment to the hilt. His favorite game was blackjack, because with a prodigious memory he assured me he had the odds over the bank if he played long enough and invested enough. So far as I could tell he was right; he certainly tried. He was a wealthy man who successfully gambled on the Stock Exchange and in the art and housing market, and his idea of relaxation (not mine!) was to go to Las Vegas. Once in Reno, by an extraordinary whim, I put a quarter into a poker machine and won $5, so I am $4.75 ahead.

Sondheimer's great chemical contribution was to the experimental side of macrocyclic polyunsaturated compounds and aromaticity. He gambled a great personal investment on his eventual ability to handle a laboriously made final compound that could turn into charcoal above –30 °C.

Great personal tragedy entered his life with the death of his beloved step-daughter Tami (a friend of my daughter). He eventually said of his work, "I'm just dotting i's and crossing t's; I don't know where I'm going." "Frank, you are in all the textbooks, what more do you want?" He had been a German–Jewish refugee and was later torn between various countries: Israel, Britain, the United States, and Mexico. He finally called quits the game of life, and I still grieve.

Disruption of life-styles appears to be important for achievement. Many creative organic chemists have backgrounds that are foreign to their present countries: Djerassi from Central Europe, Stork from Belgium and France, Sondheimer from Frankfurt, Rosenkranz from Hungary and Switzerland, Zaffaroni from Uruguay, Nakanishi from Japan and Egypt, Eliel from Cologne, Ruzicka, Prelog, and Werner from Croatia, and so forth. Perhaps this cutting of roots (including mine for a time) helps to induce an ability to be at the same time a participant and an outside observer of the game.

Editor: *From your international experience, can you perceive national as well as personal versions of chemistry?*

Science is one of the few genuinely international activities, and many of its recent developments have been made possible by airplane and later by advancing communications. Common interests and mutual understandings transcend nationality, culture, race, and creed. Nevertheless, there are some national characteristics in the pursuit of science, as expected in a cultural activity. The reasons are complex, and I can only quote a few illustrative examples.

I assume that the distribution of human talent and creativity is not racially determined. Therefore, social attitudes and interests, education systems, and material backgrounds must define the national numbers and levels available for scientists. There are problems. I recall having to teach students in Nigeria how electrical equipment and Bunsen burners work (they cooked on wood fires, with oil lamps) and to distinguish colors for which they had no words. The national attitude toward science as an intellectual activity and as a social force is more favorable in Japan, Switzerland, the United States, and the former Soviet Union than it is in Australia. The situation in the United Kingdom is ambivalent: a cultural respect for science, but a social and political suspicion of it and of its applications. In the Soviet Union there was a great respect for application (Lenin said "Socialism equals Soviets plus electricity") but a neglect of the individualism required for fundamental innovations.

France

Since 1938 I have enjoyed French civilization and culture. It is interesting to examine French organic chemistry. C. A. Wurtz, beginning his *History of Chemical Theory* in 1868, said, "*La chimie est une science Française.*" (Chemistry is a French science.) Before the First World War, the work of organic chemists like V. Grignard, P. Sabatier, and L. Claisen tended to support that statement. The bloodletting of that war, and ultimately the educational system originally established by Napoleon, reduced organic chemistry to a subsidiary science to physics and mathematics. The attitude at l'École Normale Supérieure, for example, was to exalt the intellectual status of fundamental theoretical science at the expense of the pragmatic aspect. I recall a discussion in 1957 during which we were asked in Paris to name some great French organic chemists. Apart from the obvious ones, V. Malaprade was mentioned: "*Malaprade, qui est Malaprade?*" ("Who is Malaprade?") The academics had never heard of him, but as the inventor of the periodate cleavage of diols in sugars he had made a fundamental contribution to structures in many areas, including nucleic acids. But he was and remained a demonstrator at the provincial center of Nancy, because he did not have a "respectable" academic background.

Between the wars the kind of organic research in France was exemplified by P. Cornubert, who systematically made every conceivable type of substituted cyclohexane derivative without having an inkling of the importance of conformation. Empiricism was lacking except in the chemical industry, which had to earn its living by its applications, and in the strategic research of the Centre National de la Recherche Scientifique (CNRS). World-scale fundamental advances were made in those quarters. Some of the best academic organic chemistry came from places like l'École de Pharmacie Gallenique (e.g., M. M. Janot, R. Goutarel, and J. Le Men) simply because experimental work there had an appropriate prestige. For a long time some of the outstanding academic organic chemists (e.g., Guy Ourisson and M. Lederer) were educated or highly influenced abroad. Recently the prospect has greatly changed, with trailblazing practical–theoretical achievements like that for which J.-M. Lehn was recently awarded a Nobel Prize.

India

A number of Indian collaborators have been important to my work. I know the situation well after numerous visits, one very recently, and Ph.D. thesis examinations. The British did a good job in setting up universities and other institutes. Most of the early chemical teachers and researchers were British, like J. L. Simonsen (Bangalore) or were British-trained, so the early attitudes often showed the exaggerated strengths and weaknesses of their approach. Attitudes and mechanisms are now, like Indian English, being remolded in an indigenous and more relevant fashion. A main characteristic of of Indian organic chemistry was its assiduity, often, though not always, well directed.

I had many very good Indian collaborators, one of whom (G. Nadamuni), for example, by tireless investigation, solved the difficult problem of the anomalous metal–ammonia reductions of biphenyl, which had defeated some of my British students. He is now successful in the pharmaceutical industry.

The level of organic chemistry in theory and practice is now comparable to the best in the wold, as evidenced by the work, among many others, of Ganugapati S. R. Subba Rao and G. Mehta (Hyderabad) and the recently published book on stereochemistry by my former collaborator D. Nasipuri.

Research in the drug industry is proceeding along the most modern lines. I have found also among students a diligent enthusiasm for the subject, reminiscent of the mostly lost attitudes of my youth. They do not see it just as a means of earning a living secondary to topics like medicine. Recent Indian governments are encouraging science and technology, with entrepreneurial attitudes to risk-taking. I see India as a rapidly developing powerhouse of Asia. Its levels are much more uneven than those of China (which I also know well), but it reaches greater individualistic heights.

United States

For a long time since the war, it might almost have been true to say that organic chemistry was an American science. Some supreme work has been done there. I have mentioned Robert B. Woodward, Gilbert Stork, E. J. Corey, Carl Djerassi, and W. S. Johnson. I could name many others, together with much pot-boiling. The often very useful pot-boiling arises from the desirable multivalent but time-limited funding opportunities, allied to the publish-or-perish syndrome. Nobody who does not make mistakes makes anything of value, and this applies to funding committees. The more committees there are, as in the United States, to make different mistakes (and some correct decisions), the better.

The publication–promotion–support system, prominent in the United States and elsewhere, does lead to some pursuit and publication of trivialities. (But who is to judge what is trivial?) I disagree with the Citation Index approach to assessment of work, which is influential with grants and promotions, and which appears superficially to be an unbiased "bibliometric" method. However, it tends to be used without thinking by many of the sort of people who find themselves on appropriate committees (who can count, if nothing else). If you are far ahead in a new field, you are not quoted and probably later forgotten. The typical approach by researchers is "That is so novel I am not sure that it is sound" (no quote) or "I do not understand that" (no quote) and later, "That is not new" (also no quote). Somewhat pedestrian contributions are quoted ad nauseum, like analytical methods that threaten nobody's originality and thus are routinely used. I note, genuinely without personal bias, that a named reaction like the Birch reduction is usually not referenced at all because this is considered unnecessary. On one occasion when I suggested to a very senior scientist that he should quote a predecessor in a paper, his incredible reply was "Why should I quote him? Am I not as clever as he? Could I not have thought of the idea myself?" This may not be quite as egotistical as it appears, because he certainly had done so.

The Soviet Union and Other Socialist Countries

I do not know exactly what is happening in 1995, and I am now an Academician of the new Russian Academy, not the nonexistent U.S.S.R. Academy. My friend I. V. Torgov is jubilant: "I may have no shoes, but there are also no GULAGS." (He spent a year in one because his son had scribbled on a picture of Lenin.)

As an honored outsider, I occasionally felt that there were some practical advantages to the Socialist system, but only until I hit the snags. Torgov's minute apartment, in an appalling concrete building, was as comfortable as he and his wife could make it. Nevertheless, I regarded it as totally unsatisfactory for a scientist of his international reputation and achievements. In contrast, some of the politically acceptable Akademiki who had contributed far less to the world but were involved in the Soviet public spendor enjoyed whole floors of apartment blocks in a park. At one time his office was a table at the end of a spectrometer laboratory, with nowhere to put papers and with workers bustling in and out. At a major Mendeleev Conference, part of which I chaired in Alma Ata, I publicly protested about this. Whether for that reason, or more likely because Yuri Ovchinnikov saw the light, his lot improved. He later proudly wore the red-gold badge of Hero of Soviet Labor.

One incident shows that even the Soviet police could be human. I paid a visit with Torgov in an Academy car, as Russians, to the monastery complex of Zagorsk, with its gilded domes, embroidered copes, gold reliquaries and icons, black-clothed worshipers, and deep-voiced chanting. At a road crossing we were stopped by a policeman, who was very angry for a reason I could not determine. (Perhaps the driver had not stopped, or perhaps the cop was peevish?) As usual, I started asking questions in English, to emphasize my status as a visitor.

The ploy had worked on similar occasions in Czechoslovakia and Poland, but not here. Torgov, in his gentle way, starting producing cards with his photograph, Akademia Nauk, etc., arguing quietly but with no result. Finally, the policeman burst out laughing, slapped the driver on the back, and let him go. I asked, "What did you say?" "I told him the driver had become a father for the first time that day, so the policeman said not to do it again!" I could not find out if the remark applied to the baby or to the traffic offense. The point was that, as an Academy driver, he would have lost his privileged job for the slightest infraction.

In Zagorsk, most of the worshipers were women. I asked Torgov (a Christian), "Are all Christians in Russia women?" "No, but they have more courage than men."

Why have Socialist countries not been more successful in creative science, given the enormous resources devoted to research? As a Foreign Academician, I had great privileges in the Soviet Union and access to information. In the People's Republic of China, as a Delegate of UNESCO–UNDP with funding and prestige, I enjoyed an equal access, so I have a first-hand view. The central Akademia Nauk, S.S.S.R., employed over one million people, with one Academician per million of Soviet population, not to mention other republics; for example, the

At the Presidium, Akademia Nauk S.S.S.R. in Moscow in 1976 when I was inducted as Foreign Academician. A. E. Alexandrov, then President of the U.S.S.R. Academy of Science, is at the head of the table.

Ukrainian Academy (which alone employed 70,000 scientists) or even the Kazakh Academy (which employed 30,000; nearly five times CSIRO in Australia). Scientists acceptable to the State had enormous prestige and privileges. I have also frequently visited most of the other European Socialist–Marxist countries and observed their similar systems. The answer to my question is failure attributable to human inefficiency and lack of motivation within a pyramidal system unsuited to genuine creativity. My friend Ovchinnikov said to me some years ago, "Since the Revolution there are still Czars and icons; only the faces have changed." Despite the material resources devoted, individual expressions of creativity have been restricted by the organizational structure.

Until recently (I asked about 1986), only two of the directors of the hundreds of U.S.S.R. Academy Institutes (mostly replacing industrial and government research) were in their 50s, the rest were older. Scientists worked in very large groups under close direction from the top down (god–directors) without being able to exercise individual initiative. It was a reasonable system to exploit the known, but a very bad one to produce genuine novelty. Problems, if solved at all, were usually overwhelmed by numbers in politically defined areas, like space research. I was not puzzled about why it was usually difficult to find anybody in laboratories after about 3:30 p.m.

However, as one example of genuine novelty, the breakthrough synthetic reaction that allowed industrial total synthesis of sex hormones, including oral contraceptives, was devised by I. V. Torgov and S. N. Ananchenko in a corner of a laboratory inside the Soviet Union, almost without being noticed. The Soviet Union did not benefit. The situation now appears to be changing rapidly, with encouragement of basic research, of youth, and of horizontal organization, but recovery may be slow. The world of science and technology will, however, shortly need to take real note of the results of redeployment of resources and creativities, made on the authentic basis of how research has to work. The same change is happening in the People's Republic of China, from my personal knowledge and influence as a UNESCO commissioner. As in other countries, emphases away from defense science will make resources available for more productive efforts.

Switzerland

Switzerland is an interesting country that I knew well from both academic and industrial viewpoints. My first personal acquaintance was through my friend Richard Martin (later Université Libre de Bruxelles), of the Suisse romande (Genève). For its size, it is immensely effective.

Among friends in China during a UNDP tour in 1980. The UNDP goal was to assist universities in science and engineering; later law and economics were added to the agenda.

There is no false distinction between "pure" and "applied" chemistry; there is chemistry and its applications. Many firms have academic consultants and even directors. In the past, the highest levels of the Swiss chemical industry contained chemists–academics like O. Isler, P. Plattner, A. Stoll, L. Ruzicka, and V. Prelog. Nowadays they are being replaced by lawyers and accountants. It will be interesting to see what such changes will do to its effectiveness. The internal Swiss multinational character (German, Italian, and French) is an asset.

One academic highly involved overall was Leopold Ruzicka (like Vlado Prelog, originally Croatian), part of the historical succession of émigrés (including A. Werner and R. Willstätter) that so fertilized Swiss chemistry. He had enormous personal influence and respect. He was very creative in a highly practical sense, pioneering with Taddeus Reichstein the industrial partial synthesis of sex hormones. Ruzicka embodied many contradictions. I recall asking him how he reconciled being a communist, a millionaire, and a member of the Pontifical Academy of Science. "My boy, if I lived in the Soviet Union, I would be a capitalist!" Although he was color-blind, he collected Dutch paintings.

David Craig (center), Ray Martin (right), and I at the Research School of Chemistry around 1978.

Now they can be seen as the *Stiftung Ruzicka* (Ruzicka Bequest) in the *Kunsthalle* (art gallery) in Zurich. He tried to teach me to like his favorite drink: Campari and soda, without success.

I recall a small personal illustrative episode. I was going, with Jessie, to a meeting of the Swiss Chemical Society in Glarus. Leopold had a new Mercedes and, to show it off, offered us a lift. We arrived on a gorgeous day. He stopped outside the hall and said, "You don't really want to attend dull meetings on a day like this." "Well, that is what I came for." "I thought not; let us go to the Kloenthalersee." After some highly creative driving (which to his chagrin included brushing a gate-post) we had a memorable time at that beautiful little lake. Leopold showed us cliffs he had climbed and we lunched on *Bundnerfleisch* (air-dried meat), while listening to him talking to the peasants in their dialect. He was proud of his ability in this respect; every part of Switzerland has a different dialect. On return to the hall, he happened to usher me onto the platform behind the last speaker, who was talking (in High German) about arming our brave Swiss boys with atom bombs. Leopold stood still, and so did I. Then he said loudly, in Swiss German, what I was told can be politely rendered as "What a load of nonsense", and marched out. I was left to slink in cowardly fashion into a seat. He was very happy at his rival Karrer's funeral; he had outlived him!

Ruzicka had an Alpine garden, to which he used to rush during unduly cold weather to breathe warm air on the plants. But he was in some ways a ruthless politician. He and other Swiss like Reichstein (affectionately called by his students steinreich, that is, filthy rich) were important not only for their science, but for being among the first to show that exploitation of fundamental academic science could not only assist industry, but be profitable to academics and to their research work. Ruzicka and Reichstein held many of the early patents on the partial synthesis of sex hormones and on perfumes.

He was also a protagonist in the iniquitous *pli cacheté* (sealed envelope) system, especially in contests in the commercially important area of perfumes and flavors. Then it was possible to deposit with the French or Swiss Chemical Societies a sealed and dated envelope containing scientific results that could subsequently be opened on demand, even many years later, to establish priorities between people like Y.-R. Naves and P. Karrer. The battle of the plis was simultaneously amusing and dismaying: "My pli of 23/2/32 says...." "But mine of 29/11/31 says in anticipation...." The system has been abolished. Unlike patents, it unduly inhibited open publication, but it could be profitable to individuals.

Like other entrepreneurial chemists—I can name Todd and Heilbron in the United Kingdom and Roger Adams in the United States—who founded "dynasties", Ruzicka was very good at choosing people. He also excelled at encouraging them if they met his criteria.

In my view, Swiss chemistry is of the very highest quality, combining great theoretical insight with assiduous and very capable experimentation. It is taken very seriously, often with an eye to exploitation. Nevertheless, some work, like the synthesis of vitamin B_{12} by Albert Eschenmoser in collaboration with Robert B. Woodward, was carried out only to demonstrate the power of method and imagination. Duilio Arigoni, as an international chemist although native Swiss, has a high combination of Swiss and international characteristics (including language) that place him in the highest class in the world.

Australia

I liked Australian students and collaborators. It was not just because I could understand their speech, because many English and Americans also favor them for two good reasons. The Australian university (Scottish, not U.K.) system involves a broad three-year pass degree. This program is followed, for a chosen proportion only, by a fourth year honors degree that normally involves research and a thesis. This system picks the elite on the right grounds and places much less emphasis on the "crystal ball" I had to use in Manchester. I also believe that the entrepreneurial social background of Australians was appropriate, although this is fading, with students now emphasizing rights and privileges rather than obligations. Australia produces about 2% of the refereed scientific papers in the world, with about 0.5% of the population.

Like all whites here, I am an "immigrant". In my youth Australians had few rigid paradigms, social or scientific, but we shared a belief in independence and in self-starting. In this age of plastic cards and social services, that attitude is fading, and I doubt that the next generation will feel the same. I was born an internal alien. I remember my mother, somewhat unsuccessfully, trying to inhibit my "Orstrylian" accent (which I regard now as ugly and inelegant) and my use of the Australian English language as incorrect and imprecise. (These characteristics have been developed into an art by our politicians). Speech is a reflection of thought organization. I had hoped that an enlarged "foreign" background (Australia has the highest percentage in the world of new immigrants) would help to develop a feeling for meaning in our language, but instead the result seems to be a degradation of pronunciation. As an "old Australian", I can barely understand it now. However, the prominent names in school and university results are often Vietnamese, Greek, and Italian, so perhaps a new wave is about to break.

Editor: *You have hinted that there can be less pleasant aspects in establishing one's scientific reputation and prominence. Please comment.*

In the 18th and 19th centuries, the means of publication was an exchange of private letters between workers, to assist one another. Nowadays there is instead an attitude that I label "I thought of it first". The sociology of science includes two ways of dealing with deadly rivals of about equal stature. One is to beat the rival down by whatever means can be found. If this is clearly not going to work, an opposite tack is to be taken. Build up that rival with fulsome praise, and by any means available infer an equation of the rival's achievements with yours. Thus you indirectly praise yourself and in addition show what a great person you are. This process is not often consciously calculated. Mostly it is expressed orally at conferences or in lectures. Such manipulations in the written record might be too permanent and eventually challenged.

How did this system based on personal reputation develop? The bitter prestige struggles in science exist because all else flows from reputation: positions, pay, research funding, medals, number of students, laboratory space, and perhaps even self-esteem, although that never worried me. This situation is ethically regrettable because all scientists are supported on the shoulders of their predecessors. Such struggles are also found in business and politics, but in those areas the rewards are power and money in much larger quanta. Personal prestige is more taken for granted or equated with pay.

Gamesmanship can be a rather harmless ego-building pastime, but it has a serious aspect that can affect the progress of science and especially the history of ideas. Although science deals in perceived truth and its greatest sin is faking scientific results, the presentation of the history of ideas leading to important presentations of ideas is frequently dishonest and calculated to draw the major personal reward from the system, which gives the palm to apparent novelty.

The derivative is therefore often dressed up as a breakthrough. A title like "A New Novel Reaction Carried out for the First Time Within the Precincts of Greater Miami" is a mild caricature (stolen from Gilbert Stork) of how some results are presented. Selective presentation of the literature permits you to avoid attribution to others of a fundamental idea (for example, by emphasizing one rather minor derivative aspect). This approach may be subconscious or respectably disguised as a "brief logical" presentation. A less heinous personal sin, often demanded by editors, is an economic, logical, selective presentation of facts that distorts not only previous history but what you actually did. This is a great pity, because it obscures the authentic intellectual pathways taken to reach ideas in a great human activity.

The reward system has been largely wrong, in any case. The credit for major advances in organic chemistry has usually gone to the first to determine a structure or to complete a synthesis, no matter how inelegant or impractical the method may have been. The system discourages others from finding more elegant methods and ideas. Being chronologically first in a competitive race is thought to confer automatic superiority. I have reviewed some aspects of deceit in science.[148]

I was in Oxford during much of the Robinson–Ingold and Robinson–Woodward misunderstandings (a word I select with care). These disagreements led me to realize how people could see the same things through different eyes, blinded by unscientific passions. Nevertheless, in the end the results benefit science, because challenges to established ideas determine the truth. Robinson had the broad scope of ideas. He could visualize molecules reacting in three dimensions and, having done so, he assumed he had explained a process. Ingold had a much more critically defined approach. He only accepted a mechanism if he could describe it in quantitative mathematical terms. Robinson used to comment, correctly, on much of Ingold's work, "But *I* thought of that in 1933" (or whenever). Ingold neglected Robinson altogether in his attributions. Neither could assign credit to the other for complementary approaches and achievement. It was a great pity for both as scientists and as human beings. Robinson tended to believe that subjects belonged to him permanently, so he did not have to hurry in publication, but if his discoveries were anticipated he became very upset. Indeed, I was told by a very celebrated English organic chemist that one major reason he chose his specific field was because he knew Robinson had no interest in it.

The Robinson–Ingold battle never was concluded, because neither really accepted the other's viewpoint as valid. I had one slight brush with Ingold, who wrote me an angry letter about my failure to acknowledge his contribution to the theory of reductions. In fact, his publication was not pertinent to my work. He was concerned only with reductions of protonated carbonyls through oxonium cations formed in strongly acidic solutions (Clemmensen conditions). Conversely, he never bothered to mention my initiatory ideas on reversible and irreversible protonations involved in defining the nature of products from mesomeric systems; he invented the terms "kinetic" and "thermodynamic" control instead. I never bothered to take up that matter with him. Like many other ideas in science, his theories probably evolved in parallel with mine (by seeing the obvious).

It is necessary, of course, to bring out adequately in publications the degree of novelty and scope of application. If nobody perceives the significance of your work, it is usually your own fault, and it impedes the progress of the science. I have not performed well in this connection; I assumed that predictable outcomes were as obvious to others as to me.

Editor: *Many people these days equate chemistry with environmental pollution and with often-unknown toxic dangers. What is your comment?*

The human world is full of risk-versus-benefit choices, like a car ride or a shark at Bondi beach. Of course there are problems, and precautions must be taken in both cases (the former is far more dangerous). I can only comment that both natural (e.g., the risk of getting cancer by eating celery) and unnatural (e.g., the risk of getting cancer through frying bacon) dangers have to be balanced against quality of life. It is simple to compare the general quality of human life in the 1500s in Europe (or now in many Third-World countries) with that accessible in Europe, North America, and Australia today. The favorable changes result from science in operation through technology, medicine, and organization.

But new dangers should be set against major (often now unrecognized) old ones, to be controlled by a more general public understanding of science and its potential in application. I, among many others, have tried to inculcate this understanding. The balance of "science-generated" terror is still, in my opinion, much in favor of human life and happiness, as evidenced by longevity. There will always be idiots and selfish humans, whose unethical uses of science and its derivatives (including medicinal drugs) must be politically and socially controlled. I am sorry that DuPont has had to suppress its motto: "A Better Life Through Chemistry".

***Editor:** You knew the late Robert Maxwell, and you were a member of the editorial advisory boards of a number of commercial journals. Would you like to comment on the place of commercial publication in science, and on the influence of Maxwell's Pergamon Press upon this?*

Derek Barton could answer better than I can, but I shall try. I was not happy with commercial publication in contrast to that undertaken by societies, for several reasons. Obviously, publishers have to make a profit (or they go bankrupt). As an organizer of university libraries, faced with increasing numbers and costs of journals to be obtained with decreasing funding, I had to find the extra funds and make increasingly difficult choices.

I was not sure that scientific standards in commercial journals could be maintained against profits (as they are automatically by publication committees of societies, and by their disinterested choices of referees and assessments). Against such doubts were several factors. One is that important specialized topics (e.g., organometallic chemistry in organic synthesis, or asymmetry) or multidisciplinary topics (like phytochemistry) did not initially have societies powerful enough to conduct what I regarded as desirable trans- or novel-disciplinary publications to link many artificially separated approaches to broad problem topics within one publication. The old national society publications did not aim to publish transnational work as, for example, Maxwell's journals did. Also, most discipline-based societies (like the Chemical Society, London, or the Societé Chimique, Paris, or the American Chemical Society) are national organizations that only rather incidentally publish international work (e.g., Australian papers in JCS). This situation left a big knowledge gap. Between the 1920s and the 1970s the socialist world was largely omitted from the international community of science publication. (I tried, rather unsuccessfully, to learn Russian; I can buy a meal or order a taxi!) When invited (indirectly) by Bob Maxwell to join the advisory boards of journals like *Phytochemistry* or *Tetrahedron*, I agreed on the grounds that this was the only protection that the scientist or the consumer had. Also, I approved of his basic ideas on multi- and transdisciplinary journals with a full international scope, including the socialist world (which he knew very well). In short, according to the old adage "If you can't lick 'em, join 'em!"

I could quote many examples of protective activities, but shall confine myself to one. Bob Maxwell's favorite commercial approach was to start a journal at a low price (commercial loss). When sufficient libraries were "hooked", he would raise this price with celerity. He told me once that if he could sell 2000 copies of any journal he could make a profit. He could, of course, make good arguments in favor of the procedure. ("Do you want it started at all? I am running the risks!") That I could appreciate, but we usually never knew what the real financial situation was. On one occasion he proposed a major price rise for *Phytochemistry* on the basis of "increasing expenses" (not specified in detail). The Editorial Board, on perusing the Articles of Association, discovered that the journal belonged to the Board and not to the publisher (I imagine for tax reasons). So we wrote to Bob, saying, in effect, "The journal belongs to us, so we are canceling the publication contract and seeking another publisher." He was a realist and replied to the effect, "I will keep the price at the present level if you promise to be good boys, maintaining standards and increasing subscribers." What happened later I do not know. I resigned because of my changing scientific interests, but it was and is a very successful and useful journal.

I was a member of the advisory committee to the first genuinely international chemical journal, *Tetrahedron,* from its beginning until recently, when it was taken over by Elsevier. That deal was a great shock to me, as I knew how much Pergamon Press meant to Bob; its sale must mark a grave financial crisis, I thought. I was right, but at least he ensured that it would survive. He started with scientific and technical publication and was, I believe, very emotionally involved in it. His basic theme was international communications, including the most recent computer technology. I admired him greatly in many ways, including his ability with languages (he spoke nine languages, and I could personally judge his French, English, and German as pluperfect). Unlike some other failed tycoons, he at least left something of value.[149]

Stories about *Tetrahedron*

I shall recount the "hundred golden guineas affair" as an example of Bob Maxwell's business approach to "unbribable" academics. At an IUPAC meeting in New York, Bob (who attended such affairs to meet scientists) was in a group with about half a dozen prominent organic chemists including Carl Djerassi, who was on the board of the *Journal of Organic Chemistry.* This journal had been founded by Morris Kharasch largely to take his papers that had been refused (unjustly, I am sure) by the *Journal of the American Chemical Society.* It had just been taken over by the American Chemical Society. Carl, perhaps rather idly I assume, said, "If each one of us here (sweeping around the highly distinguished group) published one of our best papers in *Tetrahedron* each year, I am sure that its circulation would surpass that of the JOC within 3 years." Bob Maxwell said, "Is that a bet? I bet each of you a hundred golden guineas that that will not be achieved." This bet (archaically aristocratic ic in its terms) was a clear inducement to publish in his new journal. In fact, although the result probably was not so influenced, the circulation was achieved within the time set. Various people then demanded their publicly negotiated dividend. Some, because of their elevation, had no difficulty. For others the negotiation process was prolonged. Carl organized his in characteristic fashion by buying a picture by Norma Redpath in London and sending the bill to Bob: "Because you owe me $*X*, I am sure you will have no hesitation in settling this account." The last person on the list was lucky, despite a prolonged wait, because the price of gold went from about $35 to $600 per ounce, and the calculation was finally US $10,000. I was not one of those involved.

Another *Tetrahedron* story concerns the medal and prize for creative organic chemistry. The first award was to M. Nesmyanov. I was told that Maxwell, who had commercial interests in the Soviet Union, wanted this award. Accordingly, he rang one influential member (Dr. *X*) of the awarding committee and said, "Dr. *Y* (another very influential member) favors Nesmyanov." He then rang Dr. *Y* and said, "Dr. *X* prefers Nesmyanov." So it was arranged. I happened to be in England, staying at the Royal Society, of which Todd was then President, and was asked to the award ceremony at a dinner in the Atheneum. (It was an awful dinner, but they had been dining there for 250 years!) During the dinner, to which Nesmyanov had been dragged (he was fatally ill) with several Russian helpers at Lord Todd's official invitation, Bob Maxwell passed a note to Todd saying, "I am very sorry, but I have forgotten to have the medal made." Todd was rightly furious, and to retrieve the situation asked Sir David Martin (across the road at the Royal Society) to produce on the spot a parchment that said, in effect, "IOU one gold medal." Everyone in sight was asked to sign, including me. I said to Bob, "I assume that this does not commit me to help fund it?"

The last time I saw Maxwell was in 1986, when he invited me to luncheon in his beautiful mansion on Headington Hill, Oxford. The conversation was mainly in French; his very pleasant wife Elisabeth comes from Arles. He asked me where I was going. "To see Alex Todd in Cambridge." "How are you traveling?" "I have hired a car." "Cancel it. I'll lend you a limousine and a chauffeur to take you anywhere in Britain, on one condition." "What it that?" "That you offer Todd a lift in it." I accepted his offer, and I did ask Todd (who then had a broken ankle) to luncheon at the Garden House, and he refused Maxwell's transport. Maxwell was on his way to Moscow (helicopter on the lawn) to produce an "English version of *Pravda*." "I thought there was already a translation?" "I am not speaking of a translation which is so dull nobody reads it, but a *version*" "I assume that you will write it yourself?" (His Russian was perfect, I am told.) At that luncheon I complained about the lack of access to research grants in Australia (because I am "too old") and explained scientifically why I needed one. He said, "You have it." He then told me that the whole of the Maxwell fortunes would go to the Pergamon Foundation. His children had had educations and jobs (which he had not had), and now it was up to them, without his fortune. Despite later unanswered requests on my part, his promise was not made good. One favored image was that of a fairy godfather, but often he could not maintain it, or perhaps forgot.

Bob Maxwell had what some people might describe as a cynical attitude. I might more favorably describe it as a realistic–political attitude. He had, after all, spent his youth as a partisan in a dirty and dangerous war. I can quote an example. At a party, during an election campaign, he looked at his watch: "I must go; I came in the Rolls, but I cannot tour my (East-end London, Labor) electorate in that. I must go home and get the Bentley." (My comment: "What is the difference?")

I never received anything from him except a few meals and some free journals, but his charm, intelligence, and attentiveness were manifest and flattering. I think he was genuinely respectful of science and scientists. I think his influence was good for international communications in chemistry, if not for library finances. I am sorry that he financially overreached himself in his personal "legend-making".

Other journals, like the *Japanese Journal of Antibiotics* with which I have been associated, were not commercial in the same sense. They were interdisciplinary publications that relied for industrial support on the prestige of an international academic–industrial board, so I believed I was contributing by helping to focus support on a worthy cause (as well as by translating papers into idiomatic English).

Editor: *Has the role of women in science changed, in your experience, over the years?*

I have known, over my 60-year scientific career, few prestigious women scientists like Dorothy Hodgkin. There were few chemists in my time in Oxford. One, Helen Muir FRS, is now director of a biological institute. Rita Harradence (Lady Cornforth) was my rival as a student in Sydney. I had very good reason to appreciate fully her intelligence and great practical skill, which certainly contributed later to her husband's Nobel prize. With three children, she abandoned her independent research (although still working) and was content to be seen as a contributor. That change summarizes a major dilemma of scientific women in the past, and a common solution.

I have had a number of very bright women students who have gone on to successful but not brilliant careers. Are women temperamentally less suited to creative science than men? Do social expectations for women inhibit their success as scientists? From where I stand, the answer is "yes" to both. This conclusion is not pejorative of women or of their present social status. The single-minded devotion (obsession) over many years that is required for high achievement in a scientific career does inhibit many of the gentler, human, aspects of life, which few women (rightly) are willing to give up. I believe that a testosterone factor, involving aggression and egotism, is needed for success at the very highest levels. This conclusion applies as well in creative fields like art and music, where women, on the whole, are not eminent as creators.

It is not surprising that many women (including my daughter Rosemary, who has high ability and is a divorced mother of two children) are content to live as human beings and to contribute scientifically without the stresses and human neglects associated with reaching the heights. My late lamented assistant Maureen Kaye, for example, by her very human and capable handling of many of my organizational responsibilities, enormously contributed to my scientific achievements, which she must notionally share. She had scientific training, but she was never afraid to "lose face" on my behalf and that of science, as I often was.

Society in the past has badly treated women, the childbearers and minders, the sole homemakers. They were seen as undeserving of higher education, but now conditions are improving. How is this change likely to affect women in science? Incidentally, in my university career I never encountered prejudice against women as such, rather the reverse. The usual problem was the lack of women with the certificated qualifications. ("Is ability being confused with documentation?" I often asked.) I still find that few women are willing to ascend into the cold, brilliant empyrean and take the personal gambles required not only for themselves but for those they love. With education and with financial and social independence, I am still not sure how many women will aim high. Statistics in favor of men might merely indicate continuing social limitations, or they might indicate that women are superior human beings.

Questions to be approached in the modern world are With the integrated nature of science and the known greater ability of women to be cooperative, is the team approach more suited to their incorporation? I can only answer yes. With the overwhelming problems of the environment (human population, pollution, conservation, exhaustion of resources), do feminine housekeeping attitudes need to be dominant? My answer is yes. Is there, in short, a new "feminine science"? I answer yes. This would involve exploitation of the known to satisfy correctly defined human needs, rather than the insatiable search for the unknown, in which I have indulged. Humanly, I approve; temperamentally, I regret. However, we must meaningfully involve both halves of the human race in the effort, on terms under which each can uniquely contribute.

Editor: *What have been the attitude changes over time in matters like laboratory design and safety?*

Traditionally, encounters with chemistry start with books, then with practical demonstrations; the encounter of the third kind is the personal handling of chemicals and equipment. I believe that some practical experiments should come first, with considered precautions. When I was 10 I dissolved a shilling (then silver) in nitric acid and got the solution on my hands, which became sequentially yellow, black, and sore. I never forgot that lesson in handling chemicals. Nowadays, safety in laboratories is often apparently more important than actually doing anything, but on looking back I am horrified at the lack of safety precautions in my early laboratories.

In my time safety regulations seemed nonexistent in the Dyson Perrins (DP) Laboratory, Oxford. Laboratory glassware was routinely cleaned in porcelain baths of boiling sulfuric–nitric acids, which was effective. But it also ruined the hoods and the slate benches. I recall a flask coated with azide (not mine) exploding; the trousers of the worker subsided in sticky folds to the floor. We threw a firebucket of water over him, sent the girls out, and he was fortunately as good as new. The problem was partly laboratory design and partly the low standard of upkeep. The open drains under the DP floorboards had a tendency to erupt in flames because there were no rules about disposal of solvents or of the metallic sodium often used to dry solvents like ether or benzene. One major fire started when a light fitting fell from the ceiling because the uninspected screws corroded. I learned lessons on management and design for practical application when I became a laboratory organizer.

When I, as the first Dean, set up the Research School of Chemistry, my first action was to issue extensive safety regulations based, through Manchester, on ICI. They were much ahead of their time and have been continuously updated. The RSC laboratory was also constructed with a close eye to safety (e.g., no "peninsular" benches, two exits to each laboratory, special hazard areas even for spectrometers, and so forth). There is no such thing as an accident, except being hit by a meteorite. With forethought and calculated precautions, accidents should not happen; some human failure is to blame.

I had only one potentially dangerous accident in my career (my own fault), despite handling very large quantities of liquid ammonia, a supposedly dangerous solvent. I regularly diluted a reduction mixture with water and extracted with ether by swirling in a conical flask. One day I thoughtlessly did this by shaking in a stoppered separating funnel, and I had not added enough water. High gas pressure blew out the stopper, and the solution (concentrated aqueous ammonia and sodium hydroxide) hit me in the face. My breathing was paralyzed and I collapsed. Fortunately, the imperturbable Renée Jaeger poured water over me and took me in hand. I lost some skin from face and mouth, but I learned several general lessons: to wear face masks or safety spectacles (which were not then provided in Oxford), to handle all reactions with knowledge and thoughtful respect, and not to become too familiar with hazards, no matter how often they are repeated. All workers in a laboratory should be informed about what is going on and what to do if there is a problem.

There is a scandalous photograph showing me doing a Birch reduction in an open hood, smoking a cigar, without safety glasses, and watched in alarm by G. S. R. Subba Rao. All I can say is that I did not know they had a camera handy (but by then I knew exactly what to do in case of a problem). I gave up smoking many years ago.

An explosion in Cambridge with ruthenium tetroxide, given to me by Todd to assess what it could do chemically, was more like a genuine accident. I dissolved it in chlorobenzene, which seemed reasonably safe, on a balance in a hood with the window down. A brown solution formed and the next instant was not there. The balance case was punched full of holes by flying glass. Todd heard the explosion from several corridors away. Alan Johnson, the only other person present, was protruding backward from under a bench. (It was not long after the war and he had a reflex reaction.) I was undamaged; I had taken precautions because of the toxic nature of the substance. I am sorry, however, that the incident put me off this interesting reagent for life.

Some reactions are unpredictably violent, and I kept clean sand to mop up and extract the products (according to Frank Lions). The best yield I ever had in a Skraup reaction was largely scraped from the ceiling, and a stain in Laboratory 21 of the Dyson Perrins bore witness for about 15 years.

Ganugapati S. R. Subba Rao in 1978 at the Dyson Perrins Laboratory, watching in some alarm while I carry out a Birch reaction and break three of my own safety rules simultaneously. The open hood, cigar, lack of safety glasses, and lack of safety precautions in general were typical of the time and place. But I never had an accident.

Editor: *To what extent are scientists driven by the desire to prove they are right? How do you deal with this psychological situation in chemistry?*

Of course, egotistical scientists (including me!) like their ideas to be proved right through their experiments. This expectation is one of the stimuli of speculation. But the whole basis of science, as opposed to superstition or religion, is dependence on experimental supports. Those scientists who design their experiments to prove that they are *right*, in my view, misunderstand the true scientific method. Experiments must be designed so that favorite ideas can be proved wrong, if indeed they are wrong, and the result must be accepted. This requirement implies that experiments should be devised to demonstrate the truth, whatever it is, perhaps regrettably free from the human emotions that often generate the original speculations. Any published erroneous conclusions must be corrected with valid experimental evidence, although scientific editors often tend to avoid such published rebuttals as polemics rather than science. After confirmation comes joy!

Editor: *I have heard you described as "the grandfather of the contraceptive pill". How do you feel about that?*

Very good, although it was one of those accidents to which science is prone. Overpopulation, "people pollution", is a staggering problem for the world's future. Any contribution toward its solution, no matter how small, is vital. I am happy also that my synthetic work has helped to liberate women in many countries, although undesirable social, religious, and economic factors have counteracted this effect in some places. It came too late for us (with five children): not that we now have regrets.

As I have recounted, contraception was not my objective nor, according to his own account, was it a goal of Djerassi. Perhaps Frank Colton, who worked for Searle, was more consciously motivated, because Gregory Pincus was a consultant of the Searle Company, which actually had the first "pill" on sale. My collaborator Herchel Smith made a major advance (a "minipill") later in Manchester. None of this could have happened when it did, or possibly for many years, without the Birch reduction, so I am happy. I draw general inferences from this regarding applications of scientific creativity, to which politicians (Australian at any rate) will not listen.

Editor: *Many of our readers are interested in the effects on spouses and families of scientific ambitions reflected in careers. In your case it seems that your ambitions and talents were responsible for moving your family not only from place to place, but across the world several times. In your retirement, how do you honestly feel about this action and its consequences?*

First, I must acknowledge that I have always had a personal sense of destiny, somehow derived from outside myself and attested by the travail I encountered in fulfilling it. If I had neglected it I would have felt like Milton: "that one talent which is death to hide, Lodg'd with me useless, though my soul more bent to serve therewith my Maker...." Because I acknowledge no Maker, perhaps the parable of the talents is more relevant. Anyway, a great sense of obligation about my gifts had to be balanced against many other personal feelings. For example, I hated air travel but accepted it, *ad nauseum,* as an obligation. Jessie recently complained that she had not, during one period of four years after retirement, coaxed me onto an international airplane.

On family matters, you should ask the members, but I think they are not unhappy. They have had many opportunities because of our actions, along with the problems. I did not lightly consider a permanent return to Australia. I regarded it, perhaps mistakenly, as a much better place for children than the United Kingdom or the United States (where I was offered attractive posts). I feel Australian, and my family members are now predominantly naturalized as such, at their choice. I recall, when leaving Australia for Britain in 1955, having a virtual nervous breakdown on the ship. I did not want to go, but my inescapable destiny called. I plead *mea culpa* to some unintentional outrages on family feelings, but *kismet*, it is written. The world, and my grandchildren, I hope, will reap some advantages. I "did it my way", egocentrically, but I am sensitive to the rights and feelings of others, particularly of those close to me.

I have had a fantastic life, one that would not have been predicted for a boy of three or so lying on the hot pavement in the slums of Paddington, Sydney, wondering why the molten black tar exuded such a strange smell. May God, if He exists, render a similar deliverance to others.

Editor: *You seem to have very wide interests inside and outside science. Is that not a handicap in scientific achievement?*

The answer depends on what you mean by achievement. A friend recently, contemplating eight of my nine bright and beautiful grandchildren, said, "Arthur, you are a rich man!" That is true in many other ways also, although as the world counts wealth I am not rich. I abhor those scientists who sit like an oyster on a rock, without a pearl, gradually swelling scientifically and financially.

Grandad and Emma enjoying the water at Coast House, New South Wales.

I was fortunate that nobody, particularly my parents, tried to make me conform to any accepted pattern. (But was that because their genes resisted it?) I think I rather astounded them, but they accepted my eccentricities. From about the age of 10 I had an intense broad curiosity concerning art, literature, music, poetry, archaeology, philosophy, ideas, languages, travel, and science. Indeed, I investigated everything except body-contact sport. At about 15 I absorbed *The Story of My Heart* by that neglected author–archaeologist Richard Jeffries, with his feeling that "soul" permeates the animate and inanimate worlds. That philosophy is one reason why I was never a "dry-as-dust" scientist.

One of my aunts prophetically christened me Professor at the age of 10 when a friend and I were constructing the *Woollahra Encyclopaedia.* The enterprise, done as a new venture on Australian "paper-bark", did not proceed much beyond Archaeology, Architecture, and Art. For me there have never been any barriers of classification of knowledge, and no distinction between a science and an art, except that for science Nature is the referee. Arts are subjective in assessment of their virtues and therefore in many ways they are more stressful to pursue. (Jessie and perhaps Carl Djerassi could comment.)

I soon recognized my own capabilities and limitations in carving my path through the natural jungle. A pivotal asset, in both science and administration, is realism and an ability to follow through a whole line of reasoning, seeing a particular piece of the obvious as part of a much larger related whole.

From 1938, when I moved to England, I was stunned by art and by history. The oldest building I had ever seen, before visiting the Pyramids, was built in 1820; it was only 8 years older than my grandfather's house at Lake Tiberias. I began to understand that symbolic ways of expressing ideas and emotions can on occasion be more effective than direct scientific expressions of reality, extensions of the evocative words of the familiar poetry that I loved. In the National Gallery and the Tate Gallery, my initial favorites were Leonardo's Virgin of the Rocks (that vision of a human world of peace and kindness and tranquillity set against the hard, unyielding, but beautiful background of nature) and the works of Turner (a feeling about, not a representation of, nature). This book was written to the music of Mozart, Bach, Beethoven, Schubert, and their ilk.

Beauty in Science

I have been driven in choosing individual scientific topics as much by emotionally perceived beauty (but how is that defined?) as by possible use or profit. I have been principally interested in discovery (what is there) rather than invention (what to do with it for humanity). I have been egotistical enough to risk taking my own emotionally appealing new lines with teaching, research, administration, and even with the architecture of laboratories, without bothering much with accepted paradigms. (It might be said that I was too lazy to acquire them.) It is a great pleasure to be proved right, but the reward is in the aesthetics of the concepts, not in material things.

If you are highly entrepreneurial, achievements can involve large sums of money. Money can be useful in doing what you want, personally and scientifically. But in my case, and for many others I am sure, the rewards are emotional. The same ecstasy accompanies the results of what I have done in science and reading what others have done, that I feel in listening to Mozart or Bach; or in reading the poetry of Shakespeare, Goethe, or Heine; in regarding Picasso's Guernica or the Sistine Chapel; or bathing in the surf of the Pacific Ocean. The reward of the true scientist is not primarily utility, not praise, not money, but the feeling of being a part of human advance into the fascinating unknown, notably of being the first human being to feel some of the hitherto unseen complexities and beauties of the universe. Of course, practical rewards do reinforce feelings of worth and make it easier for long-suffering families.

Editor: *Why do you like writing?*

I do not like general writing, any more than I like writing scientific papers, in contrast to doing the work. But my objective is much the same. Painful as it is to me to write, I believe that humanity progresses, as in science, with one generation's accomplishments based on the experiences of the previous ones. If developments are not recorded, each generation must reinvent the wheel. Many of my friends are now dead, and with each of them a personal world of experience was lost. I hope to glean my brain a little and to leave some useful food for thought based on my experiences. This may also represent a hope for spurious immortality, but I am convinced that my long and varied life and experience have given me insights that I should make known in my old age, when the results no longer matter to me personally. I love reading autobiography, so I am making some.

Editor: *You have traveled widely. Do you enjoy that as a reward of chemistry?*

Yes, indeed. I belong to the real international working jet set, not to that rather pitiable bored set seeking brief distractions. I have traveled about 2.5 million miles in 40 years (air and sea chiefly, some train). I have been in most countries in the world, except South America, and have been personally interactive with local citizens. This of course was made possible mainly by the airplane, and I have been very lucky at this juncture of world technology. In Australia, when I was a student, the shortest time to Melbourne from Sydney was 17 hours; now it is less than 2 hours.

One of the more agreeable scientific motivations was thus made a social one, as I became a member of a close-knit self-chosen international group. I have given some names from the groups of my era. As a genuinely international activity with modern communications, chemistry has some curious overtones. What is whispered in Boston one day is known in Sydney the next. Its world consists of friends and rivals, who are often identical. It is episodic in personal contacts, but relationships can be revived without apparent break after years. Many scientists have more friends outside than inside their immediate environment (as I do). Common interests and mutual understandings transcend culture, race, and creed. This "Invisible World College" is one of the great attractions of science and a great facilitator of its progress. Most scientists value the good opinion of those within the "World College", whom they respect and with whom they maintain contacts.

Editor: *You nowhere mention being ill yourself. Is good health necessary for good science?*

It certainly helps. I had 12 days of sick leave in 12 years at the ANU, and virtually no vacations (they were spent on other activities, like various inquiries, conferences, and consulting). Most scientists I have known have been very fit, as evidenced by the old age of many. (It can be an embarrassment for academies when their fellows refuse to die at the statistical rate!) I have previously been genuinely ill only once in my life (after I retired) and in the hospital for a short time. Now, at 79, I have been paying some of the debt I owe to being human. My temperament led me to foresee and avoid accidents, and my father's views on nutrition 60 or 70 years ago, which I adopted, turned out to be modern and valid. Smoking is a hazard, and at least three chemist friends have died from lung cancer. I was unwise enough to smoke myself. I blame an American GI, who gave me a carton of cigarettes that I did not know what to do with. But I gave it up many years ago. I knew of the hazards early. Our Manchester University Chancellor the 2nd Earl Woolton introduced me to Sir Richard Doll (who first publicized the cancer correlation) as "the greatest enemy of the British middle class. If he has his way, income tax will rise two and sixpence in the pound!" (This joking reference was to the high tobacco tax, but income tax rose anyway).

Chemical hazards in research laboratories, where usages continually change and proper safety regulations are observed, have had few untoward results. In contrast, workers in industry are often heedlessly exposed continuously to the same substances. I always took great care, even with the commonest substances. Despite the supposed dangers, many chemists live to a ripe old age; some are still working effectively in their 80s. I know personally, and sadly, of a few friends who have died as the result of chemical contamination: Ted Geissman (cobalt) and John Mills (paper chromatogram solvents). Of course, the greatest respect must be paid to the handling of chemical substances that present unknown, as well as known, dangers. I still have large creases in my fingers from continuous washing in my laboratory days. The safest thing to do is nothing, a situation that can be seen approaching in the United Kingdom because of unduly restrictive regulations. Of course, the scientific result is also nothing.

I have known very few people among my laboratory connections who have suffered personal health hazards. Up to the age of 76, after about 60 years in laboratories, I had not. My own remedied bladder cancer may be a result of Robinson's decision to use *o*-tolidine for steroid analogs. Its dangers were not then known. The chief problem with laboratory chemists seems to be the ready availability of means for suicide. Sadly, I have known four friends (E. Braude, Ken Pausacker, George Kenner, and Franz Sondheimer) who might be alive if this means had not been handy.

My career may have been affected by the allergies common to my family. A major motive in leaving England for Australia was to escape the grass pollen in June and July (examination season), which to me was much more serious than just a nuisance. I am allergic to few chemicals except arsenic compounds, and I cannot even be in the same laboratory with them. Some of my students have been allergic to ethylenediamine or dinitrophenylhydrazine, but I have also known people who were allergic to strawberries.

Editor: *Having seen in your lifetime the accumulation of more than half of our knowledge of chemistry (through the primitive age, the heroic age, and the instrumental age), would you like to hazard a guess as to its future?*

All that I can be sure about is that one cannot predict. In the *New Monthly Magazine* of 1814, there is an alleged news item of the year 2014: "The new steam coach proceeded from London to Halifax at a speed of 30 m.p.h. with 10 inside and 15 outside passengers." So much for the vision of transport 200 years ahead. Predictions, even by great men like Verne and Wells, tend to be extrapolations of the existing situation: bigger and better, but fundamentally the same. A very few, like Leonardo with inventions, and William Higgins (1789) with atomic and chemical reaction theory, were so far ahead of their time that their ideas had no influence. But science continues to outrun even human imagination.

Science subjects are reintegrating so that a key discovery in astronomy results in an invention in chemistry. The matrix is too large, the rate of change too rapid, and the people involved are too numerous for one person to see far ahead. Meanwhile, research committees continue to construct camels instead of horses. Nobel prizes for chemistry are now awarded mostly for biology. Does that mean biology is becoming a branch of chemistry?

Over the past 50 years it has been said that there is little left to do in organic chemistry. This attitude usually marks inflexible people who see their favorite topic, which they "know all about" grinding down. I only wish I could start all over again as an organometallic chemist, with biochemical connections. The problems expand with the broad scientific horizons of all science, particularly biology, as chemical methods, ideas, and instrumental techniques evolve to solve more and more sophisticated interdisciplinary problems. Clearly the aims and methods of organic chemistry must change continuously, giving each new generation a chance to enjoy itself and contribute in its own special way: *Mais, plus ça change plus ça reste la même chose!* (The more things change, the more they stay the same.) What will matter is how problems are defined and by whom; women will contribute their unique viewpoints. Solving problems is now easy, and needs are paramount; seeing the obvious is required for formulating the problems.

I shall consider only two broad aspects. One is that theoretical methods are developing to the point where they can fully explain and predict certain aspects of chemistry in practice. That area will continue to expand as computers and programs become increasingly powerful, and as confidence and ideas increase. I tried to develop this area (through appointing Leo Radom) at the ANU. Theoretical ideas and computers certainly have unique advantages in biochemistry (enzyme centers) and pharmacology (fits of drugs into centers).

On another front, synthetic chemistry is likely to remain an art for a long time: at least, I hope it will. Computers may greatly assist the labor involved by digesting and regurgitating information, but I do not believe they are likely to replace the Storks, Woodwards, Coreys, and so forth of this world ("rubbish in, rubbish out"). They have dangers, because they depend on fallible indexers and authors and their insights. I often used to find that the journal paper next to the one I was looking up proved the more stimulating, or that an author (and the abstracter) had no idea of the importance of what was being reported.

The future of synthetic chemistry in the laboratory lies, I believe, in the development of methods to handle, as a set of fingers would, the billions of reacting molecules as three-dimensional entities capable of specific orientation and activation. Crown ethers and related agents, highly specific polymers, the subtle uses of conformational analysis, the coordination of functional groups with metal atoms and ions (particularly the transition group, as I discussed in Research Set 4), and soluble and solid catalysts of very specific rational molecular design are examples that represent developing approaches to this common end. Ultimately laboratory synthesis should be as specific as enzyme synthesis, free of its major practical limitations (what I have styled inorganic enzyme chemistry). The periodic system is taking over organic chemistry and will continue to do so. Exactly how this will occur depends on the creativity of thousands of workers.

Editor: *What advice do you give to young scientists?*

When anyone asks me for advice, I start by asking questions. Do you really know what the pursuit of science involves? Why are you thinking of choosing it? Replies vary, including, "I have been reading the *Scientific American*, so I know all about the double helix." (In that instance I could elicit no knowledge of the fundamental importance of hydrogen bonding.) "I would like to find a cure for cancer." "Right. There is your bench, and you have access to any facilities you want. Where will you start?" "Oh, I expect you to tell me!" Or "I have been reading Djerassi's autobiography, which inspires me." "But are you sure that you are another Djerassi?" They seem to believe that each private soldier automatically has a Marshal's baton in the knapsack. This is a source of motivation, but is it based on reality? Is disappointment inevitable?

Frankly, I am not very encouraging. I ask, "Do you really know yourself at your age? Do you realize the kind of monkish/nunish existence you may need to live throughout the beginning of your career? Do you have the inner motivation and the strength to survive disappointments and defeats? Does the feeling of achievement in understanding the universe thrill you emotionally at a level equal to that of any other human activity?" But I also point out the long-term satisfactions, like international travel and interactions with people throughout the whole world, evident long-term usefulness to humanity in many ways, great emotional fulfillment through knowing, and being part of human history. In all of this, my emphasis on scientific knowledge is smaller than on their reasons for wanting to acquire and use it. Knowledge is necessary, but the motives for acquiring it are all-important.

I also ask, "If it turns out that you cannot reach the heights, can you be satisfied with a humanly useful (if less prestigious) role?" I point out that, unlike many artists and musicians who are not very useful (except to themselves), virtually all scientists and technologists can make practical and useful contributions at whatever level they reach, while earning a reasonable living.

If they are still serious despite this battering, I try to advise them on practical approaches. Science is reintegrating. You can no longer think of yourself as an organic or an inorganic chemist, or even as a chemist in relation to biological problems. A chemist needs to know enough physics to be able to handle and interpret physical techniques. You need to know about basic biology. Briefly, you cannot afford to specialize too early, although in the end your particular hard-won expertise may make an irreplaceable contribution to a common goal. You are ready to specialize when you see the whole picture and what you can add to it. I do not invite abandonment of projects too early, but rather advise beginners to "do your own thing", make your own mistakes and triumphs, and not depend on supervisors. Above all, I tell them to regard science as a unity, without any barriers that could technically exclude them.

A practical problem is that much of the necessary sacrifice occurs at an early age. Only later do the benefits accrue. How many people in these days of consumerism, nontenure posts, babies, mortgages, and social distractions can dedicate themselves for the duration? Women are prominent among those who understand the situation. Is my advice to avoid science as a career? Yes, unless you are either prepared to accept it like any conventional method of earning a living, or you have the requisite long-term dedication. I would give the same advice to an artist or a musician.

Scientific ability can only define itself in practice, so examination results are uncertain indicators either way. You need to be ready to accept risks, but the prize is great. Good luck!

(ENVOI) or "On Y'r Way, Maite" (in Colloquial Australian)

Is the Chemical Age of Legends and Heroes one with the Greeks? Does the future lie with machines and the masses, not with individual dreams and inspirations? I hope and believe not.

Further Acknowledgments

After my marriage in 1948, Sir Robert Robinson used to say in surprised tones "Your wife has done an excellent job with you." That, I acknowledge, is true. I wish to express my great admiration and affection for him, despite any remarks I have made to indicate that he was a human being. I am professionally and personally grateful to him for keeping me as an annual pensioner from 1941 to 1945, and for acting as a referee whose opinions were critical for appointments: a scholarship of the Royal Commission for the Exhibition of 1851 (in 1938), an ICI Fellowship in Oxford (1945), a Royal Society Smithson Fellowship in Cambridge (1948), the Chair in Sydney (1952), and election to the Royal Society (1958). I thank Lord Todd for his support for the Chair in Sydney, and for his research support in Cambridge. I acknowledge that I also owe to him some key attitudes in science policy. I acknowledge the personal friendship of many chemists like Leo Ellinger, Stan Slater, Alex Kemp, Serge David, Jean Crum, the Cornforths, I. R. C. and Dzunia Bick, and a host of others.

Of course, I could not have carried out alone, either physically or conceptually, the body of work recorded here. I, and organic chemistry, owe much to my dedicated, exciting, intelligent, creative, and experimentally skilled collaborators and technicians, with whom continuous interactions occurred with egos barred. I have mentioned some of my scientific collaborators, but this does not not circumscribe my gratitude to all the others. I have been greatly privileged as a focus for a ferment of ideas and activities.

I thank my son Christopher, who overcame my fear of computers in helping me to set up this document.

References

These are mostly personal, aimed to chronicle the origins of ideas and their testing, rather than to be a full literature summary to which they are an historical entry. A complete set of 444 classified publications is available from me on request. A summary by R. W. Rickards of some achievements, "political" and "scientific", is in *Festschrift Aust. J. Chem.* **1992,** *45(1),* 1, and *Chem. Aust.* **1994,** *61,* 548. As part of my receipt of the Tetrahedron Award for Creativity in Chemistry, a biography and listing of my publications up to the date of that publication can be found in *Tetrahedron* (1988, volume 44, number 10, pages v–xviii).

1. Birch, A. J. *Chem. Aust.* **1994,** *61(5),* 252.
2. Birch, A. J. *J. Proc. R. Soc. N. S. W.* **1976,** *109,* 151; *Notes Rec. R. Soc. London* **1993,** *47,* 277.
3. Birch, A. J.; Robinson, R. *J. Chem. Soc.* **1942,** 488.
4. Cardwell, H. M. E.; Cornforth, J. W.; Duff, S. R; Holterman H.; Robinson, R. *J. Chem. Soc.* **1953,** 361.
5. Birch, A. J.; Harrison, R. J. *Aust. J. Chem.* **1955,** *8,* 519.
6. Birch, A. J.; Cameron, D. W.; Rickards, R. W. *J. Chem. Soc.* **1960,** 4395.
7. Birch, A. J. Lectures 10th ICOS Bangalore 1994. In *Pure Appl. Chem.,* in press.
8. Birch, A. J. *Annu. Rep. Progr. Chem.* **1950,** 177.
9. Birch, A. J.; Walker, K. A. M. *J. Chem. Soc. C* **1966,** 1894; *Tetrahedron Lett.* **1967,** 1935, 3457; Birch, A. J.; Williamson, D. H. In *Organic Reactions;* Dauben, W. G., Ed.; Wiley: New York, 1976; p 1.
10. Birch, A. J.; Hodson, H. F.; Moore, B.; Potts, H.; Smith, G. F. *Tetrahedron Lett.* **1960,** *19,* 36.
11. Birch, A. J.; Collins, D. J.; Sultan, Muhammad; Turnbull, J. P. *J. Chem. Soc.* **1963,** 2762.
12. Birch, A. J. *J. Proc. R. Soc. N. S. W.* **1937,** *71,* 54; **1938,** *71,* 259, 261.

13. Birch, A. J.; Earl, J. C. *J. Proc. R. Soc. N. S. W.* **1938,** *71,* 330.
14. Birch, A. J.; Lions, F. *J. Proc. R. Soc. N. S. W.* **1938,** *71,* 391.
15. Birch, A. J.; Macdonald, P.; Pelter, A. *J. Chem. Soc.* **1967,** 1968.
16. Birch A. J.; Milligan, B.; Smith, E.; Speake, R. N. *J. Chem. Soc.* **1958,** 4471; Birch, A. J. *J. Chem. Soc.* **1964,** 2705.
17. Birch, A. J.; Clark-Lewis, J. W.; Robertson, A. V. *J. Chem. Soc.* **1957,** 3856.
18. Birch, A. J.; Murray, A. R. *J. Chem. Soc.* **1951,** 1888.
19. Birch, A. J.; Grimshaw, J.; Juneja, H. R. *J. Chem. Soc.* **1961,** 3126.
20. Birch, A. J.; Todd, A. R. *J. Chem. Soc.* **1952,** 3102; Birch, A. J. *J. Chem. Soc.* **1951,** 3026.
21. Birch, A. J.; Elliott, P. *Aust. J. Chem.* **1956,** *9,* 95, 238.
22. Birch, A. J. *J. Chem. Soc.* **1951,** 3026.
23. Birch, A. J.; Ritchie, E.; Speake, R. N. *J. Chem. Soc.* **1950,** 3593.
24. Birch, A. J. *J. Chem. Soc.* **1947,** 1642; Birch, A. J.; Davenport, J. B.; Ryan, A. J. *Chem. Ind.* **1956, 664.**
25. Birch, A. J.; Brown, W. V.; Corrie, J. E. T.; Moore, B. P. *J. Chem. Soc. Perkin Trans. I* **1972,** 2653.
26. Birch, A. J.; Lahey, F. N. *J. Chem. Soc.* **1953,** 379.
27. Birch, A. J.; Collins, D. J.; Muhammad, Sultan; Turnbull, J. P. *J. Chem. Soc.* **1963,** 2762.
28. Birch, A. J.; Cameron, D. W.; Rickards, R. W.; Harada, Y. *Proc. Chem. Soc.* **1960,** 22; *J. Chem. Soc.* **1962,** 303.
29. van Tamelen, E. E.; Dickie, J. P.; Loomans, M. E.; Dewey, R. S.; Strong, F. M. *J. Am. Chem. Soc.* **1961,** *83,* 1639.
30. Birch, A. J.; Musgrave, O. C.; Rickards, R. W.; Smith, H. *J. Chem. Soc.* **1959,** 3146; Birch, A. J.; Moore, B.; Rickards, R. W. *J. Chem. Soc.* **1962,** 220.
31. Birch, A. J.; Bauer, L. *Chem. Ind.* **1954,** 422.
32. Birch, A. J. *Impact: Science et Société;* UNESCO **1984,** 435.
33. Birch A. J.; Robinson, R. *J. Chem. Soc.* **1944,** 503.
34. Birch, A. J.; Collins, D. J.; Penfold, A. R. *Chem. Ind.* **1955,** 1773; Birch, A. J.; Collins, D. J.; Penfold, A. R.; Turnbull, J. P. *J. Chem. Soc.* **1963,** 792.
35. Birch, A. J.; Grimshaw, J.; Penfold, A. R.; Sheppard, N.; Speake, R. N. *J. Chem. Soc.* **1961,** 2286.
36. Budgekiewicz, H.; Djerassi, C. *J. Am. Chem. Soc.* **1962,** *84,* 1430.
37. Birch, A. J. *Chem. Ind.* **1951,** 616; *Rev. Pure Appl. Chem. Aust. Chem. Inst.* **1953,** *3,* 61.
38. Dirscherl, W. *Z. Physiol.* **1936,** *239,* 53.
39. Ehrenstein, M. *Chem. Rev.* **1948,** *42,* 457.
40. Djerassi, C. *From the Lab into the World: A Pill for People, Pets, and Bugs;* American Chemical Society: Washington, DC, 1994.

41. Birch, A. J. *J. Chem. Soc.* **1944,** 430.
42. Birch, A. J.; Jaeger, R.; Robinson, R. *J. Chem. Soc.* **1945,** 58243.
43. Snyder, H. R.; Werber, F. X. *J. Am. Chem. Soc.* **1950,** *72,* 2692, 2965.
44. Birch, A. J.; Subba Rao, G. S. R. *Aust. J. Chem.* **1970,** *23,* 547.
45. Kharasch, M.; Tawney, H. *J. Am. Chem. Soc.* **1941,** *63,* 2308.
46. Birch, A. J.; Robinson, R. *J. Chem. Soc.* **1943,** 501; Birch, A. J.; Smith, M. *Proc. Chem. Soc.* **1962,** 356.
47. Birch, A. J.; Brown, J. M.; Subba Rao, G. S. R. *J. Chem. Soc.* **1964,** 3309; **1965,** 5139; **1967,** 2509; *Tetrahedron* Suppl. 8, Part 1, pp 359, 966.
48. Birch, A. J.; Smith, H. *J. Chem. Soc.* **1951,** 1882; Birch, A. J.; Thornton, R. E.; Smith, H. *Chem. Ind.* **1956,** 1310; *J. Chem. Soc.* **1957,** 1339.
49. Fieser, L. F.; Fieser, M. *Steroids;* Reinhold: New York, 1959.
50. Cornforth, J. W.; Cornforth, R. H.; Robinson, R. *J. Chem. Soc.* **1942,** 689.
51. Campbell, B. K.; Campbell K. N. *Chem. Rev.* **1942,** *31,* 78.
52. Wooster, C. B.; Godfrey, K. L. *J. Am. Chem. Soc.* **1937,** *59,* 596.
53. Birch, A. J.; Mukherji, S. M. *J. Chem. Soc.* **1949,** 2531; Birch, A. J. *J. Chem. Soc.* **1950,** *47,* 210.
54. Ehrenstein, M. *J. Org. Chem.* **1944,** *9,* 435.
55. Birch, A. J. *Annu. Rep. Prog. Chem. London* **1951,** *48,* 184.
56. Birch, A. J. (and other authors including Djerassi, C.) In *Steroids;* Butterworth-Heinemann: Stoneham, MA, 1992; pp 57, 354, et seq.
57. Djerassi, C. *Science (Washington, D.C.)* **1966,** *151,* 1055.
58. Birch, A. J.; Graves, J. M. H.; Subba Rao, G. S. R. *J. Chem. Soc.* **1965,** 5137.
59. Birch, A. J. *Q. Rev. Chem. Soc.* **1950,** *4,* 69; Birch, A. J.; Smith, H. *Q. Rev. Chem. Soc.* **1958,** *12,* 17; Birch, A. J.; Subba Rao, G. S. R. *Adv. Org. Chem.* **1972,** *8,* 1; Smith, H. *Organic Reactions in Liquid Ammonia;* Interscience: New York, 1963; Birch, A. J. *Annu. Rep. Prog. Chem. London* **1950,** 47.
60. Birch, A. J.; Hinde, A. L.; Radom, L. *J. Am. Chem. Soc.* **1980,** *102,* 3370; 4074; 6430; **1981,** *103,* 204.
61. Birch, A. J.; Nasipuri, D. *Tetrahedron* **1959,** *6,* 148.
62. Birch, A. J. *J. Chem. Soc.* **1950,** 1551.
63. Birch, A. J. *J. Chem. Soc.* **1950,** 2325.
64. Birch, A. J.; Smith, H. *J. Chem. Soc.* **1951,** 1882.
65. Birch, A. J.; Brown, J. M.; Stansfield, F. *Chem. Ind.* **1964,** 1917; *J. Chem. Soc.* **1964,** 5343; Birch, A. J.; Graves, J. M. H.; Subba Rao, G. S. R. *J. Chem. Soc.* **1965,** 5137. Birch, A. J.; Subba Rao, G. S. R. *Tetrahedron* Suppl. 7 **1966,** 391; Birch, A. J.; Iskander, G. M.; Magboul, B. I.; Stansfield, F. *J. Chem. Soc. C* **1967,** 358; Birch, A. J.; Keeton, A. J. R. *J. Chem. Soc. C* **1968,** 109; *Aust. J. Chem.* **1971,** *24,* 513. Birch, A. J.; Graves, J. M. H.; Siddall, J. B. *J. Chem. Soc.* **1963,** 4234.

Birch, A. J.; Brown, J. M.; Subba Rao, G. S. R. *J. Chem. Soc.* **1964,** 3309; *Tetrahedron* Suppl. **1966,** *7,* 291.

66. Birch, A. J. *Rev. Pure Appl. Chem. (R. Aust. Chem. Inst.)* **1953,** *3,* 61.
67. Birch, A. J.; Powell, V. H. *Tetrahedron Lett.* **1970,** 3457.
68. Birch, A. J.; Butler, D. N.; Siddall, J. B. *J. Chem. Soc.* **1964,** 2941.
69. Birch, A. J.; Hill, J. S. *J. Chem. Soc. C* **1966,** 419.
70. Birch, A. J.; Macdonald, P. L.; Powell, V. H. *J. Chem. Soc. C* **1970,** 1469.
71. Birch, A. J.; Dastur, K. P. *J. Agric. Food Chem.* **1974,** 22, 162.
72. Birch, A. J.; Hextall, P. *Aust. J. Chem.* **1955,** *8,* 96.
73. Birch, A. J.; Wright, J. J. *Aust. J. Chem.* **1969,** 22, 2635.
74. Subba Rao, G. S. R.; Pramod, K. *Proc. Ind. Acad. Sci.* **1984,** 573. Birch, A. J.; Mani, N. S.; Subba Rao, G. S. R. *J. Chem. Soc. Perkin Trans. I* **1990,** 1423. Subba Rao, G. S. R.; Devi, L. Uma; Sheriff, Uma Javad *J. Chem. Soc. Perkin Trans. I* **1991,** 964.
75. Birch, A. J. *J. Chem. Soc.* **1945,** 809; **1947,** 102.
76. Birch, A. J.; Slobbe, J. *Heterocycles* **1976,** *109,* 151.
77. Birch, A. J.; Looker, C. T.; Madigan, R. T. Report Indep. Inquiry Commonwealth Sci. Ind. Res. Organisat. Aust. Gov. Publ. Service, Canberra, Australia, 1977.
78. Birch, A. J. *Fortschr. Chem. Org. Naturst. Springer Wien* **1957,** *14,* 186. *Proc. Chem. Soc.* **1962,** 3. Interdiscip. Sci. Rev. **1976,** *1,* 215.
79. Winterstein, E.; Trier, G. Die Alkaloide Gebrüd. Bornträger, Berlin, 1910.
80. Willstatter, R. *From My Life (Aus Meinem Leben);* Stoll, A. W. A., Ed.; Benjamin: New York, 1965; pp 235, 445.
81. Robinson, R. *J. Chem. Soc.* **1917,** 876.
82. Birch, A. J. *Notes Rec. R. Soc.* **1993,** *47,* 277.
83. Birch, A. J.; Donovan, F. W. *Aust. J. Chem.* **1953,** *6,* 360.
84. Raistrick, H. *Proc. R. Soc.* **1948,** *A199,* 141; *B136,* 481; *Suomen Kemi* **1950,** *23,* 221. Geissman, T. A.; Hinreiner, E. *Botan. Rev* **1952,** *18,* 77.
85. Collie, J. N. *J. Chem. Soc.* **1907,** *91,* 1806. Stewart, A. W. *Recent Advances in Organic Chemistry,* 4th Ed.; Longmans Green: London, 1920.
86. Robinson, R. *J. R. Soc. Arts* **1948,** *96,* 795.
87. Birch, A. J.; Donovan, F. W. *Aust. J. Chem.* **1955,** *8,* 529.
88. Nolan, T. J. In *Thorpe's Dictionary of Applied Chemistry;* Longmans Green: London, 1946, pp 7, 293.
89. Birch, A. J.; Massy-Westropp, R. A.; Moye, C. J. *Aust. J. Chem.* **1955,** *8,* 539.
90. Birch, A. J.; Elliott, P.; Penfold, A. R. *Aust. J. Chem.* **1954,** *7,* 169. Birch, A. J.; English, R. J.; Massy-Westropp, R. A.; Slaytor, M.; Smith, H. *Proc. Chem. Soc.* **1957,** *204,* 365. Birch, A. J.; Massy-Westropp, R. A.; Slaytor, M.; Smith, H. *J. Chem. Soc.* **1958,** 365.

Birch, A. J.; Cassera, A.; Fitton, P.; Holker, J. S. E.; Smith, H.; Thompson, G. A.; Whalley, W. B. *J. Chem. Soc.* **1962,** 3583.

91. Birch, A. J.; Cameron, D. W.; Holloway P. W. *Tetrahedron Lett.* **1960,** *25,* 26. Birch, A. J.; Holloway, P. W.; Rickards, R. W. *Biochem. Biophys. Acta* **1962,** *57,* 143.
92. VII Corso Estivo (1962) Chim. Accad. Naz. Lincei Roma **1964,** *57,* 67, 73.
93. Birch, A. J.; Snell, J. F.; Thompson, P. J. *J. Am. Chem. Soc.* **1960,** *82,* 2402. *J. Chem. Soc.* **1962,** 425.
94. Robinson, R. *The Structural Relations of Natural Products;* Oxford: Great Britain, 1955 (alleged account of the Weizmann Lectures of 1953).
95. Birch, A. J.; Donovan, F. W. *Aust. J. Chem.* **1953,** *6,* 373.
96. Birch, A. J.; Slaytor, M. *J. Chem. Soc.* **1961,** 4692.
97. Birch, A. J.; Massy-Westropp, R. A.; Moye, C. J. *Aust. J. Chem.* **1955,** 539. For other general summaries see Birch, A. J. In *Perspectives in Organic Chemistry;* Todd, A. R., Ed.; Interscience: London, 1956; *Science (Washington, D.C.)* **1967,** *156,* 3772; *Some Recent Developments in the Chemistry of Natural Products;* Rangaswami S.; Subba Rao, N. V., Eds.; Prentice Hall: India, 1972, p 1. *Further Perspectives in Organic Chemistry;* CIBA Symp. **1978,** *53,* 5. *Pure Appl. Chem.* **1978,** *50,* 105.
98. Rickards, R. W. quoted in Birch, A. J.; Smith, H. *Chem. Soc. Spec. Public* **1958,** *12,* 11. Birch, A. J.; Hussein, S. F.; Rickards, R. W. *J. Chem. Soc.* **1964,** 3494.
99. Birch, A. J.; Blance, G. E.; Smith, H. *J. Chem. Soc.* **1958,** 4582.
100. Birch, A. J.; English, R. J.; Massy-Westropp, R. A.; Slaytor, M.; Smith, H. *J. Chem. Soc.* **1958,** 365, 369. Birch, A. J.; Ryan, A. J.; Schofield, J.; Smith, H. *J. Chem. Soc.* **1965,** 1231.
101. Birch, A. J.; Hussein, S. F.; Rickards, R. W. *J. Chem. Soc.* **1964,** 3494.
102. Birch A. J. *Biologically Active Compounds from Peptides;* CIBA Symp. Churchill **1958,** 247. Birch, A. J.; McLoughlin. B. J.; Smith, H. *Tetrahedron Lett.* **1960,** *7,* 1. Bhattacharji, S. C.; Birch, A. J.; Brack, A.; Hofmann, A.; Kobel, H.; Smith, D. C. C.; Smith, H.; Winter, J. *J. Chem. Soc.* **1962,** 421. Birch, A. J. *Chem. Weekblad* **1968,** *56,* 597.
103. Birch, A. J.; Anet, E. F. L. J.; Massy-Westropp, R. A. *Aust. J. Chem.* **1957,** *10,* 93.
104. For refs. see Richards, J. H.; Hendrickson, J. B. *The Biosynthesis of Terpenes, Steroids, and Acetogenins;* W. A. Benjamin: New York, 1964.
105. Birch, A. J.; Kocor, M.; Sheppard, N.; Winter, J. *J. Chem. Soc.* **1962,** 1502.
106. Birch, A. J.; Rickards, R. W.; Smith, H.; Harris, A.; Whalley, W. B. *Tetrahedron* **1959,** *7,* 241.
107. Birch, A. J. *Amino Acids and Peptides with Antimetabolic Activity;* CIBA Symp. Churchill **1958,** *247,* 257. Birch, A. J.; McLoughlin, B. J.;

Winter, J. J. *Chem. Ind.* **1960,** 840. Birch, A. J.; Allen, M.; Jones, A. R. *Aust. J. Chem.* **1970,** 23, 427.

108. Birch A. J.; Moye, C. J. *J. Chem. Soc.* **1957,** 412; **1958,** 2622.
109. Birch A. J.; Slaytor M. *Chem. Ind.* **1956,** 1524.
110. Birch, A. J.; Maung, M.; Pelter, A. *Aust. J. Chem.* **1969,** 22, 1923.
111. Birch, A. J.; Holzapfel, C. W.; Rickards, R. W.; Djerassi, C.; Suzuki, M.; Seidel, P. C.; Westley, J.; Dutcher, J. D.; Thomas, R. *Tetrahedron Lett.* **1964,** *1485,* 1491.
112. Quilico, A.; Piozzi, F.; Cardani, C. *Gazz. Chim. Ital.* **1958,** *88,* 125.
113. Birch, A. J.; Blance, G. E.; David S.; Smith, H. *J. Chem. Soc.* **1961,** 3128. Birch, A. J.; Farrar, K. R. *J. Chem. Soc.* **1963,** 4277.
114. Birch, A. J.; Wright J. J. *Tetrahedron* **1970,** *26,* 2329. Birch, A. J.; Baldas, J.; Russell, R. A. *J. Chem. Soc. Perkin Trans. I* **1974,** 50.
115. Birch, A. J.; Cameron, D. W.; Holzapfel, C. W.; Rickards, R. W. *Chem. Ind.* **1963,** 379. Birch, A. J.; Holzapfel, C. W.; Rickards, R. W. *Tetrahedron* Suppl. 8, Part II **1966,** 359.
116. Birch, A. J.; Effenberger, R.; Rickards, R. W.; Simpson, T. J. *Tetrahedron Lett.* **1976,** 2371; *J. Chem. Soc. Perkin Trans. I* **1979,** 807, 816.
117. Kogl, F.; Wessem G. C.; Elsbach, O. L. *Rec. Trav. Chim.* **1945,** *84,* 23.
118. Birch, A. J.; Cassera, A.; Rickards, R. W. *Chem. Ind.* **1961,** 792.
119. Birch, A. J. *Pure Appl. Chem. (IUPAC)* **1963,** *7,* 527. Additional work to be published.
120. Birch, A. J. 6th. International Biochemistry Conference abstracts, 1962. Birch, A. J.; Cameron, D. W.; Holloway, P. W.; Rickards, R. W. *Tetrahedron Lett.* **1960,** 25, 26.
121. Birch, A. J.; Grimshaw, J.; Turnbull J. P. *J. Chem. Soc.* **1963,** 2412.
122. Birch, A. J. *Proc. XVII IUPAC Munich Pure Appl. Chem. 1959;* Butterworth: London, 1960, p 73. *Chemistry of Flavonoid Compounds;* Geissman, T. A., Ed.; Macmillan: New York, 1962, p 623.
123. Lawrence, W. C. J.; Price J. R. *Biol. Rev. Cambridge Philos. Soc.* **1940,** *15,* 35.
124. Birch, A. J.; Cross, P. E.; Lewis, J.; White, D. A. *Chem. Ind.* **1964,** 838. *J. Chem. Soc. A* **1968,** 332; Birch A. J.; Chamberlain, K. G. B.; Haas, M. A.; Thompson, D. J. *J. Chem. Soc. Perkin Trans. 1* **1973,** 1882. Birch, A. J.; Williamson, D. H. *J. Chem. Soc. Perkin Trans. I* **1973,** 1900.
125. Birch, A. J.; Jenkins, I. D. *Transition Metal Organometallics in Organic Synthesis;* Vol. I; Alper, H., Ed.; Wiley: New York, 1976, p 1.
126. Birch, A. J.; Walker, K. A. M. *Tetrahedron Lett.* **1967,** *34,* 275. 1935. *Aust. J. Chem.* **1971,** *24,* 513. Birch, A. J.; Williamson, D. H. In *Organic Reactions;* Dauben, W. G., Ed.; Wiley: New York, 1976, Chapter 1.
127. Birch, A. J. *Ann. N. Y. Acad. Sci.* **1980,** 333. Birch, A. J.; Bandara, B. M. R.; Chamberlain, K. B.; Chauncy, B.; Dahler, P.; Day, A. I.; Jenkins, I. D.; Kelly, L. F.; Khor, T.-C.; Kretschmer, G.; Liepa, A. J.;

Narula, A. S.; Pearson A. J.; Raverty, M.; Rizzardo, E.; Sell, E. C.; Stephenson, G. R.; Thompson, D. J.; Williamson, D. H. *Tetrahedron* (Woodward Mem. Issue) **1981**, 37, Suppl. 9, 289. Birch, A. J. *Current Sci. (India)* **1982**, *51*, 155. Birch, A. J.; Kelly, L. F. *J. Organomet. Chem.* **1985**, *285*, 267. Birch, A. J.; Bandara, B. M. R.; Kelly, L. F.; Narula, A. S.; Raverty, W. D. *J. Org. Chem.* **1981**, *46*, 5166; *J. Chem. Soc. Perkin Trans. I* **1982**, 1745, 1755.

128. Birch, A. J.; Westerman, P. W.; Pearson, A. J. *Aust. J. Chem.* **1976**, *29*, 1671. Birch, A. J.; Bogsanyi, L. F. K. D. *J. Organomet. Chem.* **1981**, *214*, C39. Bandara, B. M. R.; Birch, A. J.; Raverty, W. D. *J. Chem. Soc. Perkin Trans. I* **1978**, 638; *J. Chem. Soc. Perkin Trans. I* **1982**, 1763.
129. Birch, A. J.; Chamberlain, K. B.; Thompson, D. J. *J. Chem. Soc. Perkin Trans. I* **1973**, 1900. Birch, A. J.; Kelly, L. F.; Narula, A. S. *Tetrahedron Lett.* **1969**, 4107; **1980**, 871. Birch, A. J.; Narula, A. S.; Dahler, P.; Stephenson, G. R.; Kelly L. F. *Tetrahedron Lett.* **1980**, 2455; **1982**, 1813.
130. Birch, A. J.; Bandara B. M. R.; Raverty, W. D. *J. Chem. Soc. Perkin Trans. I* **1982**, 1763.
131. Birch, A. J.; Cross, P. E.; Fitton, H. *Chem. Soc. Chem. Commun.* **1965**, *15*, 366.
132. Birch, A. J.; Kelly, L. F.; Thompson, D. J. *J. Organomet. Chem.* **1985**, *286*, 37.
133. Birch A. J.; Chamberlain, K. B. *Org. Synth.* **1975**, *57*, 16.
134. Birch, A. J.; Bandara, B. M. R.; Kelly, L. F.; Khor, T.-C. *Tetrahedron Lett.* **1983**, 2491. Birch, A. J.; Raverty, W. D.; Stephenson, G. R. *Organometallics* **1984**, *3*, 1075.
135. Birch, A, J.; Liepa, G. R. S. *J. Chem. Soc. Perkin Trans. I* **1982**, 713.
136. Birch, A. J.; Kelly, L. F.; Liepa, A. *Tetrahedron Lett.* **1985**, 501.
137. Birch, A. J.; Raverty, W. D.; Hsu, S. Y.; Pearson, A. J. *J. Organmet. Chem.* **1984**, *260*, C59.
138. Birch, A. J.; Narula, A.; Kelly, L. F. *Tetrahedron Lett.* **1980**, *1455*, 2981; *Tetrahedron* **1982**, *38*, 1813.
139. Bandara, B. M. R.; Birch, A. J. *J. Organomet. Chem.* **1984**, *202*, C6.
140. Birch, A. J.; Chauncy, B.; Kelly, L. F.; Thompson, D. J. *J. Organomet. Chem.* **1985**, *266*, 37.
141. Birch, A. J.; Stephenson, G. R. *Organometallics* **1984**, *3*, 1075.
142. Birch, A. J.; Cross, P. E.; Connor, D. T.; Subba Rao, G. S. R. *J. Chem. Soc.* **1966**, 54.
143. Birch, A. J.; Bandara, B. M. R.; Raverty, W. D. *J. Chem. Soc. Perkin Trans. I* **1982**, 1745.
144. Birch, A. J.; Raverty, W. D.; Stephenson, G. R. *Tetrahedron Lett.* **1980**, 197; *J. Chem. Commun.* **1980**, 857. Birch, A. J.; Narula, A. S.; Kelly, L. F. *Tetrahedron Lett.* **1891**, 779.
145. Birch, A. J.; Kelly, L. F. *J. Org. Chem.* **1984**, *49*, 2496.
146. Birch, A. J.; Kelly, L. F.; Weerasuria L. *J. Org. Chem.* **1988**, *53*, 278.

147. Birch, A. J.; Birch, M. J. *Aust. J. Chem. Resource Book;* Fogliani, C. L., Ed.; *R. Aust. Chem. Inst.* **1991,** *10,* 54.
148. Birch, A. J. *Interdiscip. Sci. Rev.* **1990,** *14,* 334; **1992,** *17,* 95.
149. Barton, D. H. R.; Baldwin, J.; Birch, A. J.; Djerassi, C.; Ourisson, G.; Thompson, H.; Wilkinson, G. et al. In *Robert Maxwell and Pergamon Press;* Pergamon: New York, 1988.

Index

Index

B

C

D

E

F

G

H

I

J

K

L

M

N

O

P

Q

R

T

U

W

X

Z

Copy editing: Janet S. Dodd
Production: Margaret J. Brown
Indexing: Janet S. Dodd

Production Manager: Cheryl Wurzbacher

Printed and bound by Maple Press, York, PA

Highlights from ACS Books

Good Laboratory Practice Standards: Applications for Field and Laboratory Studies
Edited by Willa Y. Garner, Maureen S. Barge, and James P. Ussary
ACS Professional Reference Book; 572 pp; clothbound ISBN 0–8412–2192–8

Silent Spring Revisited
Edited by Gino J. Marco, Robert M. Hollingworth, and William Durham
214 pp; clothbound ISBN 0–8412–0980–4; paperback ISBN 0–8412–0981–2

The Microkinetics of Heterogeneous Catalysis
By James A. Dumesic, Dale F. Rudd, Luis M. Aparicio, James E. Rekoske, and Andrés A. Treviño
ACS Professional Reference Book; 316 pp; clothbound ISBN 0–8412–2214–2

Helping Your Child Learn Science
By Nancy Paulu with Margery Martin; Illustrated by Margaret Scott
58 pp; paperback ISBN 0–8412–2626–1

Handbook of Chemical Property Estimation Methods
By Warren J. Lyman, William F. Reehl, and David H. Rosenblatt
960 pp; clothbound ISBN 0–8412–1761–0

Understanding Chemical Patents: A Guide for the Inventor
By John T. Maynard and Howard M. Peters
184 pp; clothbound ISBN 0–8412–1997–4; paperback ISBN 0–8412–1998–2

Spectroscopy of Polymers
By Jack L. Koenig
ACS Professional Reference Book; 328 pp;
clothbound ISBN 0–8412–1904–4; paperback ISBN 0–8412–1924–9

Harnessing Biotechnology for the 21st Century
Edited by Michael R. Ladisch and Arindam Bose
Conference Proceedings Series; 612 pp;
clothbound ISBN 0–8412–2477–3

From Caveman to Chemist: Circumstances and Achievements
By Hugh W. Salzberg
300 pp; clothbound ISBN 0–8412–1786–6; paperback ISBN 0–8412–1787–4

The Green Flame: Surviving Government Secrecy
By Andrew Dequasie
300 pp; clothbound ISBN 0–8412–1857–9

Other ACS Books

Biotechnology and Materials Science: Chemistry for the Future
Edited by Mary L. Good
160 pp; clothbound, ISBN 0–8412–1472–7, paperback, ISBN 0–8412–1473–5

Chemical Demonstrations: A Sourcebook for Teachers
Volume 1, Second Edition by Lee R. Summerlin and James L. Ealy, Jr.
192 pp; spiral bound; ISBN 0–8412–1481–6
Volume 2, Second Edition by Lee R. Summerlin, Christie L. Borgford, and Julie B. Ealy
229 pp; spiral bound; ISBN 0–8412–1535–9

The Language of Biotechnology: A Dictionary of Terms
By John M. Walker and Michael Cox
ACS Professional Reference Book; 256 pp;
clothbound, ISBN 0–8412–1489–1; paperback, ISBN 0–8412–1490–5

Cancer: The Outlaw Cell, Second Edition
Edited by Richard E. LaFond
274 pp; clothbound, ISBN 0–8412–1419–0; paperback, ISBN 0–8412–1420–4

Chemical Structure Software for Personal Computers
Edited by Daniel E. Meyer, Wendy A. Warr, and Richard A. Love
ACS Professional Reference Book; 107 pp;
clothbound, ISBN 0–8412–1538–3; paperback, ISBN 0–8412–1539–1

Practical Statistics for the Physical Sciences
By Larry L. Havlicek
ACS Professional Reference Book; 198 pp; clothbound; ISBN 0–8412–1453–0

The Basics of Technical Communicating
By B. Edward Cain
ACS Professional Reference Book; 198 pp;
clothbound, ISBN 0–8412–1451–4; paperback, ISBN 0–8412–1452–2

The ACS Style Guide: A Manual for Authors and Editors
Edited by Janet S. Dodd
264 pp; clothbound, ISBN 0–8412–0917–0; paperback, ISBN 0–8412–0943–X

Personal Computers for Scientists: A Byte at a Time
By Glenn I. Ouchi
276 pp; clothbound, ISBN 0–8412–1000–4; paperback, ISBN 0–8412–1001–2

Chemistry and Crime: From Sherlock Holmes to Today's Courtroom
Edited by Samuel M. Gerber
135 pp; clothbound, ISBN 0–8412–0784–4; paperback, ISBN 0–8412–0785–2

For further information and a free catalog of ACS books, contact:
American Chemical Society
Distribution Office, Department 225
1155 16th Street, NW, Washington, DC 20036
Telephone 800–227–5558